HELL ON ICE
THE SAGA OF THE "JEANNETTE"

HELL ON ICE

THE SAGA OF THE "JEANNETTE"

BY

COMMANDER
EDWARD ELLSBERG

WILLIAM HEINEMANN LTD
LONDON   ::   TORONTO

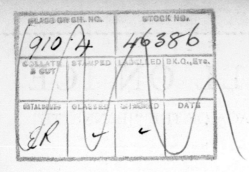
FIRST PUBLISHED 1938

PRINTED IN GREAT BRITAIN AT THE WINDMILL PRESS
KINGSWOOD, SURREY

TO

EMMA WOTTON DE LONG

STILL WAITING AFTER SIXTY YEARS TO
REJOIN THE MAN WHO SAILED AWAY
IN COMMAND OF THE "JEANNETTE"

" . . . a truer, nobler, trustier heart,
More loving or more loyal, never beat
Within a human breast."

PREFACE

ON the summit of a grassy hill in Maryland looking across an arm of the Severn River toward the spreading lawns and the grey buildings of the Naval Academy stands a stone cross frosted with marble icicles topping an oddly shaped granite cairn.

In the summer of 1910, a boy of eighteen fresh from the Colorado Rockies, I stood, a new midshipman in awkward sailor whites, before that monument and read the inscription to Lieutenant-Commander G. W. De Long and the officers and men who perished with him in the *Jeannette* Expedition of 1879 in search of the North Pole. Casually I noted that no one was buried beneath that cross, and since I had never heard before either of De Long or of the *Jeannette,* I wandered off to study the monuments to naval heroes whose deeds shone out in the histories I had read—the officers who in the wars with Tripoli had humbled the Barbary pirates; those who in the Civil War had braved Confederate forts and iron-clad rams to save the Union; and most of all to stand before the tomb of John Paul Jones, the father of our Navy and a valiant seaman, fit companion to the great commanders of all ages.

Over the next twenty years I heard again occasionally of De Long in connection with the successful expeditions to the North and to the South Poles, finally reached by Peary and by Amundsen and those who followed in their footsteps. But except as a dismal early failure, De Long's expedition seemed to have no significance, until some seven years ago a brief article by a friend of mine, Commander Louis J. Gulliver, appeared in the Naval Institute summarising so splendidly the history of the *Jeannette* that immediately that old stone cross in Annapolis for me took on a new importance and I began to study what had happened. Reading what I could get my hands on concerning it, I soon enough saw that De Long's early failure was a more brilliant chapter in human struggle and achievement than the later successes of Peary and of Amundsen.

vii

But in my early search, based mainly on De Long's journals as published nearly sixty years ago, much of what had happened eluded me; first, because De Long himself, fighting for the lives of his men in the Arctic, never had opportunity to set down in his journal what was going on (the most vivid day of his life is covered by two brief lines); and second, because the published version of his journal was much expurgated by those who edited it to create the impression that the expedition was a happy family of scientists unitedly battling the ice, whereas the truth was considerably otherwise as I soon learned.

Fortunately there came into my hands the old record of the Naval Court of Inquiry into the disaster, before which court the survivors testified, from which it appeared that De Long's struggles with his men tried his soul even as much as his struggles with the ice; and on top of that discovery, with the aid of Congressman Celler of New York, I got from the records of Congress the transcript of a Congressional Investigation lasting two solid months, a volume of nearly eleven hundred closely printed pages, from which the flesh to clothe the skeleton of De Long's journal immediately appeared. For there, fiercely fought over by the inquisitors (Congressional investigations apparently being no different over half a century ago from what they are to-day) were the stories of every survivor, whether officer or man, dragged out of him by opposing counsel, insistent even that the exact words of every controversy, profane as they might be, go down in the record to tell what really happened in three years in the ice-pack. And there also, never otherwise published, were all the suppressed reports relating to the expedition, the expurgated portions of De Long's journal, and the unpublished journals of Ambler and of Collins.

From the records of these two inquiries, Naval and Congressional, backed up by what had been published—the journal of De Long appearing as "The Voyage of the *Jeannette*"; "In the Lena Delta," by G. W. Melville, chief engineer of the expedition; and "The Narrative of the *Jeannette*," by J. W. Danenhower, navigator—stood forth an extraordinary human story. Over this material I worked three years.

How best to tell that story was a puzzle. De Long and the *Jeannette* Expedition had already most successfully been embalmed and buried by loving hands in the sketchy but conventional historical treatments of the published volumes mentioned above. To repeat that method was a waste of time. It then occurred to me that since I had once narrated in the first person in "On the Bottom" the battle of another group of seamen (of whom I was one) with the ocean for the sunken submarine S-51, I might here best give this story life and reality by relating it in fictional form as the personal narrative of one of the members of the expedition.

But who should that man be?

It was, of course, obvious that he must be chosen from the group of survivors. That narrowed the field to three officers and eight seamen. Now as between officers and seamen, it was evident that the officers were in a far better position to observe and to know what was happening than the seamen, so the choice was limited to the three surviving officers. For reasons that will afterwards be clear, among these three there could hardly be any question— Melville patently was best. And aside from the fact that Melville was a leading light in the expedition and next to De Long himself the man who actually bore the brunt of Arctic fury, he was an engineer, and since I am also, I could most easily identify myself with him and with his point of view.

So here as it might have been told about thirty years ago by Admiral George Wallace Melville, retired Engineer-in-Chief of the Navy, blunt, loyal, and lovable, a man whose versatility in four widely dissimilar fields of human endeavour gave him at his death in 1912 good claim to being considered one of America's geniuses, is the Saga of the *Jeannette*.

EDWARD ELLSBERG.

CHAPTER I

THIS year, 1909, deserves remembrance for one thing at least aside from the retirement into private life of President Roosevelt. A few weeks ago through the Virginia Capes steamed into Hampton Roads our battle-fleet, sixteen salt-crusted veterans of an unprecedented adventure—the circumnavigation of the globe by an entire fleet. There they were, back from the distant seas, guns roaring in salute to our President, flags flying everywhere, whistles from craft of all kinds shrieking them a welcome home.

Roosevelt, unafraid as always, had sent them out in the teeth of unnumbered critics who foresaw our battleships with broken-down machinery rusting in every foreign port from Valparaiso to Gibraltar, but instead with engines smoothly turning, the blunt noses of those sixteen battleships ploughed back sturdily into Hampton Roads.

I had never had any fears. I had watched the machinery of every one of those sixteen ships grow on the drafting tables of the Bureau of Steam Engineering—pistons, cranks, connecting-rods, boilers, pumps, condensers. My life went into the design of those engines and boilers on every ship, and from the flagship *Connecticut* proudly leading the long line down to the distant battleship bringing up the rear of the column, there wasn't a boiler, there wasn't a steam cylinder, that wasn't part of George Wallace Melville. Under my eyes, under my guidance, they had grown from ideas on the drawing-board to the roaring kettles and the throbbing engines before which panting coal-heavers and sweating oilers toiled below to drive those beautiful white hulls round the world and safely home to Hampton Roads.

But now I can foresee the day of those ships is done, and I think I have discernment enough left to see that mine is also. Here in this year 1909, hardly six years since my retirement as Engineer-in-Chief of the Navy, I look upon the vast fleet the machinery of which I designed, and I see its passing. Last year the *Lusitania*, turbine-driven, speeding across the Atlantic to a new record, sounded the knell of the huge reciprocating engines I designed for all those battleships. And practically completed, waiting to join her older sisters, was the *Delaware*, our newest ship, a "dreadnought" so they call her now, a huge ship of 20,000 tons, but—fired by oil! Her oil fires spell the doom of the romance of the fire-room—the stokers, the grimy coal-passers, the slice-bar—that pandemonium, that man-made inferno, with forced-draft fans roaring, with the clang of coal-buckets trolleying from bunker to fire-room floor, with the glare of the flames on sweating torsos as the furnace doors swing back and brawny arms heave in the coal! They'll all go soon, flying connecting-rods and straining coal-heavers, driven out by the prosaic turbine and the even more prosaic oil-burner.

But so it goes. We marine engineers dream, design, and build, to send forth on the oceans the most beautiful creations man turns out anywhere on land or sea—but soon our ships fade from existence like a mist before the sun. For sixteen years I was Engineer-in-Chief for the Navy, and the machinery of that battle-fleet the nation watched so proudly steaming home through the Capes was my creation, but I've seen enough in the fifty years since I entered the Navy when the Civil War broke out to doubt that ten years more will find a single ship of that armada still in active service. Turbines, oil-burning boilers, bigger guns, heavier armour—they are crowding in fast now, and soon my ships will go to the wreckers to make way in the fleet for the bigger and faster vessels sliding down the building ways in the wake of the huge *Delaware*.

Odd how one's perspective changes with the years! As a young engineer, I would have believed with those cheering thousands last month in Hampton Roads that to have had a guiding hand in creating that fleet would be the high light in my life—but now I know better. In the end it is how men lived and died, not the material things they constructed, that the world is most likely to remember. That is why in my mind a stone cross in Annapolis Cemetery looms larger and larger as the years drift by. Years ago, hewn from a driftwood spar, I set up the original of that cross in the frozen Lena Delta to stand guard over the bodies of my ship-mates; that stone replica in Annapolis, silent marker of their memory, will loom up in our history long after there has completely vanished from the seas every trace of the ships and the machinery which the world now links with the name of Melville.

We were seeking the North Pole back in 1879 when I came to set up that cross. To-day, exactly thirty years later, they're still seeking it. At this very moment, unheard from for months, Peary is working north from Greenland. I wish him luck; he's following a more promising route than that one through Behring Sea which we in the *Jeannette* found led only to disaster.

It's strange. The roar of guns in battle, machinery, boilers, hot engine-rooms and flaming fire-rooms, have made up most of my life since that day in 1861 when as a young engineer I entered the Navy to go through the Civil War, but now at sixty-eight, what sticks most in my mind is still that cruise of long ago when for two years our boiler fires were either banked or out, our engine never made a revolution, engineering went by the board, and with only the Aurora Borealis overhead to witness the struggle, with me as with all hands on the *Jeannette,* existence settled into a grim question of ice versus ship, and God help us if the ship lost!

We were an odd company there in the *Jeannette's* ward-room, five naval officers and three civilians, drawn together seeking that chimera, a passage through Behring Sea to the Pole. De Long, our captain, was responsible mainly for our being there—George Washington De Long—a man as big as his namesake, scholarly in appearance, to which a high forehead, a drooping moustache, and his glasses all contributed, but in spite of that a self-willed man, decisive, resolute, eager to be the first to end the centuries-old search for what lay at the Pole. Behind De Long in this affair was James Gordon Bennett, owner of the *New York Herald,* and an outstanding figure in American journalism. Shortly before, Bennett had won world-wide notice and acclaim for the *Herald* by sending Stanley on the seemingly hopeless task of finding Livingstone in the wilds of unknown Africa and then topped off that success by backing Stanley's amazing explorations on the Congo and the headwaters of the Nile. Bennett, seeking now fresh worlds to conquer in the interests of journalism, was easily persuaded by De Long to turn his attention and his money from conquered equatorial Africa to the undiscovered Pole. It was Bennett who purchased the *Jeannette* and put up the cash to fit her out. But once the ship was bought, Bennett hardly figured in the actual expedition. That was De Long's show from beginning to end. And what an end!

I joined the *Jeannette* as engineer officer in San Francisco in April, 1879. An uninviting wreck she looked to me then alongside the dock at the Mare Island Navy Yard, torn apart by the navy yard workmen for the strengthening of her hull and for the installation of new boilers. A chequered history the *Jeannette* had had before I ever saw her—originally as the *Pandora* of the Royal Navy; then, with guns removed, in the hands of Sir Allen Young, as a private yacht in which her owner made two cruises to high latitudes in the Arctic seas. Finally, she was bought in England from Young by Bennett on De Long's recommendation as the most suitable vessel available for the projected Polar voyage.

The most suitable she may have been—over that point experts have wrangled through the years since. So far as I am concerned, the *Jeannette* was satisfactory. But the naval constructors and engineers at Mare Island, California, when De Long after a passage round the Horn in her sailed his purchase into the Navy Yard, made no bones about saying they thought De Long had been badly fooled and the ship would scarcely do. But what they thought of the *Jeannette* was neither here nor there. Bennett had bought her, De Long was satisfied with her. The criticisms of the naval experts at Mare Island, three thousand miles away, got little attention in Washington, where with the power of the *New York Herald* behind him and De Long's enthusiasm to batter down all opposition, naval or otherwise, Bennett got a bill through Congress making the *Jeannette* a naval vessel, and (while Bennett was still to stand all the expenses of the expedition) directing the Navy to furnish the personnel and carry the project through as a naval undertaking.

So when I joined the ship there in San Francisco, I found her torn to pieces, with Lieutenant Chipp, who was to be executive officer, and Master Danenhower, slated to go as navigator, already on the spot following up the alterations as representatives of De Long. Danenhower, soon promoted to lieutenant, had joined in Havre and rounded the Horn with her. Lieutenant Chipp had shortly before arrived from China to take the post as executive officer. And during the weeks which followed my own arrival, came the others to fill out the officers' mess—Surgeon Ambler; Mr. Collins, meteorologist; Mr. Newcomb, naturalist; and Mr. Dunbar, ice-pilot. A queer collection we were, as I well learned months before De Long's dying fingers scrawled the last entry in the *Jeannette's* log, and Fate played queer tricks with us.

CHAPTER II

NATURALLY, as her engineer officer, I scanned with deep interest every detail of the vessel to which I was to trust my life in the Arctic, and I may say that torn wide open as she lay when I first saw her, I had an excellent opportunity to get intimately acquainted with the *Jeannette's* scantlings and with her machinery.

Even for that day, 1879, the *Jeannette* was a small ship, hardly 420 tons in displacement. She was only 142 feet long, 25 feet in the beam, and drew but 13 feet of water when fully loaded. She was a three-master, barque rigged, able in a fair breeze under full sail to make six knots, which, not to hold anything back, was almost two knots better than I was ever able to get her to do with her engines against even an ordinary sea.

Obviously, not having been built for Arctic service, the *Jeannette's* hull required strengthening to withstand the ice, and when I first saw her, from stem to stern the ship was a mad-house, with the shipwrights busily tearing her apart as a preliminary to reinforcing her hull and otherwise modifying her for service in the north. Amidships was a huge hole in her deck through which her original boilers, condemned by a survey, had been lifted out to be junked. To make more room for coal (for we were outfitting for a three-year cruise) the old boilers were being replaced by two smaller ones of a more efficient and compact design, by which device our coal stowage was increased in capacity nearly fifty per cent—an achievement of no mean value to a ship which, once we left Alaska, would have no opportunity to refuel on her voyage.

But this change in the fire-room, radical as it was, was trifling in comparison with the additions being made to the hull itself. To strengthen her for ramming into the ice-fields and to withstand the ice, the bow below the berth deck for a distance of ten feet abaft the stem was filled in solid with Oregon pine timbers, well bolted through and through. Outside in this vicinity, her stem was

6

sheathed with wrought iron, and from the stem back to the fore-chains, row on row laid on horizontally, a series of iron straps was bolted to the outer planking to shield it from ice damage.

In way of the boilers and engines, completely covering her side framing, the inside of the ship was sheathed fore and aft with Oregon pine planks six inches thick, extending from the boiler bed timbers up the side to the lower deck shelf; and outside the ship from just above the water-line to well below the turn of the bilge, a doubling of five inches of American elm had been added, so that the total thickness of the *Jeannette's* side when we finally sailed was over nineteen inches, a thickness which put her in the class with *Old Ironsides* when it came to resisting local penetration.

But the work did not stop there. The sides might be invulner-able locally but still collapse as a whole like a nut in a nut-cracker when gripped between two ice-floes. To resist any such contingency, in addition to the two original athwartship bulkheads which sup-ported the sides laterally, an athwartship truss of massive wood beams, 12 by 14 inches in section, braced diagonally against the bilges and the lower side of the main deck, was installed just for-ward of the new boilers to bolster the sides amidships; while just abaft these boilers there was refitted an old iron truss which the ship had previously carried somewhat farther forward. The result of these additions was that so far as human ingenuity could pro-vide, the *Jeannette* was prepared to resist both penetration and crushing in the ice. Certainly no steamer before her time had set out better braced to withstand the Arctic ice-fields.

My major interest, of course, was with the main machinery. On the *Jeannette*, this consisted of two back-acting engines, each with a thirty-two-inch diameter cylinder and an eighteen-inch stroke, developing a total of 200 horse-power at about 60 revolutions, which on our trials in the smooth waters of San Francisco Bay, gave the ship a speed of about five knots. Our shaft led aft through the stern-post to a two-bladed propeller, nine feet in diameter, so arranged under a well in the stern that the propeller could be unshipped and hoisted aboard whenever desired, which clearly enough was a valu-able feature on a vessel subjected to ice dangers.

During our fitting-out period all this machinery was carefully overhauled, four extra blades for our propeller were provided; and at my request, two new slide valves for the main engines were fitted, in order to change the cut-off and give the engines a greater expansion, which by increasing the economy of steam consumption would conserve to the utmost our precious coal.

Aside from the above there were many minor items—the addition of another auxiliary pump (a No. 4 Sewell and Cameron); the installation of a complete distillation plant to provide us with fresh water; and the fitting on deck of a hoisting and warping winch made of a pair of steam-launch engines rigged out with the necessary gearing and drums for handling lines.

Not in my department, but of interest to all hands who were going to live aboard, were the changes made to the ship itself to increase its habitability in the north. Material for a portable deck-house to cover our main deck over the forecastle was furnished us, and all exposed iron-work throughout the vessel was felted over. An entrance porch was built over the forward end of the poop, leading to the officers' quarters, and given to us in a knocked-down state, while the insides of both the forecastle and the ward-room were thickly covered with felt for insulation.

The thousand and one details in fitting out that we had to go into, I will pass over. De Long was in Washington, smoothing out difficulties, financial and otherwise, with the Navy Department, and obtaining all information on previous Polar expeditions, both foreign and domestic, on which he could possibly lay a hand. Consequently all through the spring, on Chipp, on Danenhower, and on myself at Mare Island fell the task of following up the repairs and alterations; of getting the most we could done to the ship at the least expense; and as every naval officer who has ever taken his ship through an overhaul period well knows, of battling through the daily squabbles between ship's officers and navy yard personnel as to who knew better what ought to be done and how best to do it. We did our utmost to tread on no one's toes, but from the beginning the officers at the Navy Yard regarded the *Jeannette* herself as unsuitable for a serious polar voyage, and this hardly led to complete harmony between them and us; an unfortunate situation which I think may have also been aggravated somewhat by doubts on their part about what the *Jeannette* Expedition was really intended for— a newspaper stunt for the glorification of James Gordon Bennett, or a *bona fide* attempt to add to the scientific knowledge of the world? But whatever their feelings, they did a thorough job on the ship, even though the cost, about $50,000, must have been something of a shock to Mr. Bennett, who, after paying for the repairs previously made to the *Jeannette* in England, probably felt the vessel ready to proceed to the Pole with only a perfunctory stop at Mare Island to take aboard stores and crew. And I know, especially in the beginning of this fitting-out period, that De Long himself was on tenterhooks for fear that the cost of all these unexpected repairs and replacements would cause Bennett to abandon the enterprise. He was constantly, in his letters from Washington, cautioning us to use our ingenuity and our diplomacy with the Yard's officers to affect every practicable economy, and whenever possible within the terms of the Act of Congress taking over the *Jeannette,* to see that costs, especially for materials furnished, were absorbed by the Navy itself and not lodged against the expedition.

So we struggled along through April, May, and June, with my dealings on machinery mainly with Chief Engineer Farmer of the Navy Yard, while Chipp worked with Naval Constructor Much, who handled all the hull work at Mare Island, and Danenhower confined himself to disbursing the funds and watching the accounts. The two new boilers (originally intended for the U.S.S. *Mohican* but diverted to us to expedite completion) were finally dropped into our hold, the beams and decking replaced, and the *Jeannette,* though life aboard was still a nightmare as the vessel rang from end to end under the blows of shipwrights' mauls and caulking-hammers, once more began to look something like a ship instead of a stranded derelict.

CHAPTER III

MEANWHILE our crew was being assembled, an unusual group naturally enough in view of the unusual nature of our projected voyage.

Of Lieutenant De Long, captain of the *Jeannette,* originator of the enterprise, and throughout its existence the dominant spirit in it, I have already spoken. The choice of the others who made up the expedition, especially of those ranked as officers, rested with him. Good, bad, or indifferent, they were either selected by him or met his approval; no one else was to blame if, before our adventure ended, of some he wrote in the highest terms while others were at various times under arrest by his orders, and with one at least he was engaged in a bitter feud that lasted to the death of both.

Second in command was Lieutenant Charles W. Chipp, whom I have already briefly noted in connection with the repairs. He came all the way from the China Station to join the ship as executive officer. Chipp, of moderate height, who in appearance always reminded me of General Grant, due both to his beard and his eyes, was a calm, earnest, reticent sort of person, serious, rarely given to smiles, and a first-class officer. He was an old shipmate of De Long's in the U.S.S. *Juniata,* and together they had had some previous Arctic experience when in 1873 their ship was sent north to the relief of the lost *Polaris.* On this mission, when the *Juniata,* not daring because of ice conditions to venture farther north, was stopped at Upernavik, Greenland, both De Long and Chipp cruised together for nearly two weeks in a small steam launch several hundred miles farther to the northward, searching among the bergs of Baffin Bay for the *Polaris's* crew. To their great disappointment they failed to find them, a circumstance not, however, their fault, since unknown to the searchers, the *Polaris* survivors had already been rescued by a Scotch whaler and taken to Great Britain. On this hazardous voyage, covering over 700 miles in a 33-foot steam launch, amidst the bergs and gales of Baffin Bay, Chipp as De Long's second got his baptism of ice, and in all the intervening years from that adventure, even from the distant Orient, he kept in close touch with De Long, eager if his shipmate's dreams of a polar expedition of his own ever materialised, to take part. When, early in 1878, the *Pandora* was finally purchased in England, Chipp was in China, attached to the U.S.S. *Ashuelot.* Upon learning of this concrete evidence of progress toward the Pole, he tried strenuously to secure his immediate detachment and join the renamed *Pandora* in England for the trip round the Horn, but in this he was unsuccessful, and it was not until late April, 1879, that by way of the Pacific he finally arrived from Foo Chow to join us in San Francisco.

The third and last of the line officers was Master John W. Danenhower, navigator, who a few weeks after we sailed, in the regular course of naval seniority, made his number as a lieutenant. Danenhower, the youngest of the officers aboard, having been out of the Naval Academy only eight years at the time, was during the summer of 1878 on the U.S.S. *Vandalia,* convoying ex-President Grant, then at the height of his popularity, on a triumphal tour of the Mediterranean. Here, off the coast of Asia Minor, the news of Bennett's purchase of the *Pandora* for a polar expedition reached him. Whether prompted by youthful exuberance or a desire to escape the heat of the tropics, I never knew, but at any rate, Danenhower promptly got in touch, not with De Long, but with Bennett, offering his services, and shrewdly enough backing up his application with an endorsement obtained from the *Vandalia's* distinguished passenger, General Grant himself!

For the owner of a Republican newspaper, this was more than sufficient. Bennett promptly accepted him, subject only to De Long's approval. Scheduled shortly to sail from Havre to San Francisco with the *Jeannette* and confronted with the imperative need of finding immediately, to help him work the ship, some assistant in the place of the distant Chipp who was still ineffectually struggling in China to get his detachment, De Long gladly assented to this solution. Between Bennett, General Grant, and the cables, Navy red tape was rudely cut, Danenhower's transfer swiftly arranged, and after a hasty passage by steamer and train across the Mediterranean and Europe, he arrived from Smyrna shortly before the *Jeannette* shoved off from Havre.

Danenhower, hardly thirty when we started, masked his youth (as was not uncommon in those days) behind an ample growth of side-whiskers. Unlike his two seniors in the Line, he had had no previous Arctic experience of any kind, but he was enthusiastic, impetuous, big in frame, strong and husky, and from all appearances better able than most of the rest of us to withstand the rigours of the north.

Concerning myself, then a passed assistant engineer in the Navy with the rank of lieutenant, little need be said. Of all the regular officers on the *Jeannette,* I was the oldest both in length of service and in years, being at the time we set out thirty-eight and having entered the Navy when the war began in 1861 as a third assistant engineer. My years in the poorly ventilated and hot engine-rooms of those days had cost me most of my hair, but to compensate for this, I had the longest and fullest beard aboard the *Jeannette,* which I think gave me somewhat of a patriarchal appearance to which, however, my age hardly entitled me. Oddly enough, my first Polar service was coincident with that of De Long and Chipp, for I was engineer officer of the U.S.S. *Tigress,* also searching Baffin Bay in 1873 for the *Polaris* survivors when we fell in with the *Juniata's* launch, officered by De Long and Chipp, on the same mission searching off Cape York. When the launch had exhausted its small coal supply and returned to the *Juniata* we in the *Tigress* (which was really a purchased whaler and therefore better suited for the job than a regular naval vessel like the *Juniata*) continued the task.

My first acquaintance with De Long had come, however, several years earlier than this, when in the sixties, we were shipmates on the U.S.S. *Lancaster* on the South Atlantic Station, where in spite of the fact that he was on deck and I in the engine-room, we got to know each other well. It was as a result of this friendship and the interest in polar research I had myself acquired on the *Tigress* that, at De Long's suggestion, I volunteered my services for the *Jeannette.* I had some difficulty getting the berth, however, for the Bureau of Steam Engineering, being hard pressed for personnel, was loath to let anyone in the Bureau itself go.

So far as her operation as a ship went then, these four of us, three officers of the Line, De Long, Chipp, and Danenhower, and one officer of the Engineer Corps, myself, made up the commissioned personnel of the *Jeannette*.

We had with us one more officer of the regular navy, Passed Assistant Surgeon James M. Ambler, a native of Virginia and a naval surgeon since 1874. Upon the recommendation of the senior medical officers of the Navy, Ambler was asked by De Long to take the berth, and gladly accepted.

I met Ambler for the first time on the *Jeannette*. Quiet, broad of brow, dignified in manner and bearing, of amazing vitality, he impressed me from the first both as an excellent shipmate and as a competent surgeon to whose skill, far from hospitals and resources of civilisation, we might safely trust our health in the Arctic. And in this belief, the hazards of the months to come proved we were not mistaken.

These five mentioned, regularly commissioned in the Navy, comprised the whole of those technically entitled to be considered as officers, but in the ward-room mess we had three others, Collins, Newcomb, and Dunbar, who came into that category in spite of the fact that they were shipped as seamen. The Act of Congress taking over the vessel authorised the Secretary of the Navy to detail such naval officers as could be spared and were willing to go, but as for the rest of the crew, it permitted only the enlistment of others as "seamen" for this "special service." To some degree this created a dilemma which from the beginning had in it the seeds of trouble, for as a scientific expedition, Bennett desired to send along certain civilians. These gentlemen, who obviously were not seamen, and who felt themselves entitled to consideration as officers (in which belief the rest of us willingly enough concurred) were nevertheless informed by the Navy Department that legally they could go only as "seamen for special service" or not at all. How they were to be considered aboard ship and what duties they might be assigned, would rest with the commanding officer.

This fiat of the Department, for the men concerned a bitter pill to swallow, was soon ameliorated by De Long's assurance that those affected were to be treated as officers, and on this understanding Collins, Newcomb, and Dunbar were accordingly shipped as "seamen for special service." And then and there was laid the basis of a quarrel which long after those involved were stretched cold in death, mercifully buried by the snowdrifts on the bleak tundras of the Lena Delta, still raged in all the unbridled malevolence of slander and innuendo through naval courts and the halls of Congress, venomously endeavouring to besmirch both the living and the dead.

Jerome J. Collins, of the staff of the *New York Herald,* was appointed by Mr. Bennett as meteorologist of the expedition, but was obviously aboard mainly as a newspaper-man. Collins, a big man with a flowing moustache but no beard, was active, energetic, eager in a news sense to cover the expedition, often in trouble with the rest of us, for the usual naval temperament, taught to regard the captain's word as law, was wholly missing in *this* newspaper-man's ideas of the freedom befitting a reporter. Collins's appointment as meteorologist was natural enough for he ran the weather department of the *Herald,* though scientifically his knowledge of meteorology was superficial. But he plunged whole-heartedly into the subject, and aided by De Long who got him access both to the Smithsonian Institution and to the Naval Observatory, he absorbed all he could on meteorology in the few months which elapsed while we were fitting out.

Still, shipping as a "seaman" rankled in Collins's soul; a small thing to worry over perhaps, but he was over sensitive and often took offence when none was meant. Many times since have I wondered whether Collins, the only one amongst the ward-room mess not born an American, may not, like many immigrants, have been unduly tender on that account and therefore imagined subtle insults in the most casual comments of his shipmates about his birthplace, Ireland.

Our other civilian scientist was Raymond Lee Newcomb of Salem, Massachusetts, naturalist and taxidermist of the expedition.

Newcomb, serious, slight in build, small as compared to the rest of us, seemed at first glance ill adapted to stand the gaff of a polar voyage, but technically he was a good naturalist and that settled his appointment, in spite of his boyish manner.

Last of all those comprising the ward-room mess was William Dunbar, ice-pilot, who hailed from New London and had been a whaler all his life, had commanded whalers in the Behring Sea, and of all those aboard, had had the longest and the most thorough knowledge of ice, ice-packs, and the polar seas. By far the oldest man aboard either in the ward-room or in the forecastle, Dunbar's grizzled face, grey hairs, and fund of experience gave his words on all things Arctic an air of authority none of the rest of us could muster, and on his knowledge and sagacity as ice-pilot, rested mainly our hopes of navigating the *Jeannette* safely through the ice-fields.

These were the eight that made up the *Jeannette's* ward-room mess, each in his own way looking to the ice-fields and the mysterious regions of the Pole as the path to knowledge, to adventure, or to fame. Instead, even after thirty years, my heart still aches when I recall what the ice did to us and where for most of us that path led.

CHAPTER IV

THROUGHOUT May and June we were busy loading stores, coaling ship, running our trials, cleaning up the odds and ends of our alterations, and signing on the crew.

De Long in Washington, deluged from all over the country with requests from young men, old men, cranks, and crackpots of every type, eager to go along in all sorts of ridiculous capacities, diplomatically solved his difficulty by rejecting each claim in about the same letter to all:

"I have room in the *Jeannette* for nobody but her officers and crew. These must be seamen or people with some claim to scientific usefulness, but from your letter I fail to learn that you may be classed with either party."

And then, having thus disposed of the undesirables, from Washington he wrote the *Jeannette,* carefully instructing Chipp as to the essential requirements for the seamen he desired to have signed on:

"Single men, perfect health, considerable strength, perfect temperance, cheerfulness, ability to read and write English, prime seamen of course. Norwegians, Swedes, and Danes preferred. Avoid English, Scotch, and Irish. Refuse point-blank French, Italians, and Spaniards. Pay to be Navy pay. Absolute and unhesitating obedience to every order, no matter what it may be."

De Long's instructions with respect to nationalities were based mainly on his assumptions with regard to their supposed abilities to withstand the rigours of the north, but they seemed to me to a high degree humorous when I consider that I, of Scotch descent, fell in the class to be "avoided," while De Long, himself of French Huguenot parentage, came in the group, "to be refused point-blank."

How little the average American of that day went to sea may be inferred from the fact that it did not even enter De Long's mind to mention "Americans" among the various categories to be considered for his crew, though not forty years before in the hey-day of wooden ships, the sails of Yankee clippers manned by Yankee seamen, whitened every ocean.

In early June, De Long came west and at Mare Island joined us to witness the final completion and trial of the ship. Completely satisfied with the changes made in the ship, he had nothing but praise for the manner in which we, his three subordinates, had carried on in his absence, and waved aside the gloomy prognostications of the Navy Yard officers and their comments about inadequate spars and sails, improper shape of hull, and (to put it briefly) the *Jeannette* as a whole, which they damned euphemistically in their official trial report:

"*So far as practicable,* we are of the opinion that she has been repaired and placed in condition for service in the Arctic Ocean."

But on one thing, securing an escort as far as Alaska, De Long had firmly set his heart. He was anxious to get from San Francisco into the Arctic as rapidly as possible to take advantage of what summer weather he could in working his way north. The weather at sea to be expected being mostly head-winds, speed meant proceeding under steam rather than under sail on the long trip to Alaska. This of necessity would use up most of our coal, forcing us to start the Arctic part of our journey with our bunkers either empty or what was almost as bad, full of such inferior and almost unburnable coal as was available in Alaska; unless an escort ship accompanied us as far as the Arctic Circle to replenish our bunkers then with the excellent anthracite obtainable in San Francisco.

Regardless, however, of all his arguments and his persuasions, De Long was unable to get the commandant at Mare Island to approve the detail of any naval vessel for this duty; nor, with Bennett unfortunately abroad, did he have, in spite of his most urgent telegrams, any better luck in forcing the Navy Department itself to order one. In this dilemma, at the last minute Bennett saved the situation by a cable from Paris, authorising the charter of a schooner, the *Fanny A. Hyde,* to carry the coal north. De Long, relieved of his worry but exasperated beyond measure by the controversy, eased his mind by wiring back to the *Herald,*

"Thank God, I have a man at my back to see me through when countries fail!"

On June 28, the *Jeannette* was commissioned as a ship of the Navy. Thirty years have passed since then, but still that gay scene is as fresh and bright in my memory as only yesterday. Our entire ship's company was mustered on the poop for the ceremony, officers in the glitter of swords, gold lace, and cocked hats to starboard, seamen in sober navy-blue to port. Between those lines of rough seamen about to dare the Arctic ice, Emma De Long stepped forth, as fresh and lovely on that June day as summer itself, the very embodiment of youthful feminine grace if ever I have seen it in any port on this earth. With a dazzling smile that seemed to take in not only our captain but every member of his crew as well, she manned the halliards and amidst the hoarse cheers of the sailors swiftly ran aloft our flag, a beautiful silken ensign lovingly fashioned for her husband's ship by the slender fingers which for the first time now hoisted it over us. And then with a sea-going salute as our new banner reached the mast-head, she passed the halliards to the quartermaster, stepped back to De Long's side, and clung proudly to his arm while he read the orders detailing him to the command, and the commandant of the Navy Yard, Commodore Colhoun, formally (and no doubt thankfully) turned the ship over to him.

A few days later, under our own steam we moved from Mare Island to San Francisco and there, away from the din of the yard workmen finally, we finished in peace loading stores in preparation for departure.

At last, on July 8, 1879, with the North Pole as our destination, the *Jeannette* weighed anchor, and gaily dressed out in all her signal flags, slowly steamed through the harbour, escorted by all the larger craft of the San Francisco Yacht Club, while as an indication of the esteem of Californians generally, Governor Irwin himself accompanied us to sea aboard a special tug.

That was a gala day for San Francisco, climaxing a week of banquets and farewell parties given for us in the city. Telegraph Hill was black with cheering crowds; on every merchant vessel in the bay as we passed flags dipped and impromptu salutes rang out. From the Presidio, a national salute blazed from the fortifications as we passed, the Army's god-speed on our mission to their brothers of the Navy. It was well the Army saluted us, for that was the only official salute we received on our departure. De Long, much chagrined, noted as I noted, that not a single naval vessel, not a single naval officer, took part in the ovation at our departure, and that though three warships, the *Alaska,* the *Tuscarora,* and the *Alert,* lay at the Navy Yard only twenty-six miles away, and one of them at least might have been sent for the purpose. And as if to emphasise the point, the navy yard tug *Monterey,* which that very morning had brought the commandant down to San Francisco on other business, not only lay silently at her wharf while we steamed by her, but fifteen minutes later crossed our wake hardly a mile astern of us and without even a blast of her whistle as a farewell, steamed off in the direction of Mare Island.

From the machinery hatch where I was keeping one eye out for my clumsy engines and the other out for my last long glimpse of home, I watched in puzzled surprise as the *Monterey* silently disappeared astern, then looking up at the bridge nearly overhead, caught a glimmer of a wry smile on De Long's face as he watched the *Monterey.* A navy expedition, we were sailing without the presence of a single naval representative in that vast crowd of men and ships cheering us on; worse perhaps, for it seemed as if we were being studiously ignored. Why? I have often wondered. Not enough rank on the *Jeannette,* perhaps.

But this was the only cloud on our departure, and I doubt if overwhelmed in the roar of the guns from the Presidio, the cheers from the citizens of San Francisco, and the shrieking of whistles from the flag-decorated vessels as we passed, others, especially civilians, ever noticed it.

Slowly, under engines only, we steamed out the Golden Gate and met the long swells of the Pacific. Astern, one by one the escorting yachts turned back. On the starboard wing of the bridge at her husband's side stood Emma De Long, a sailor's daughter, and after a hectic courtship terminating finally in a sudden shipboard marriage in the far-off harbour of Le Havre, for eight years now a sailor's wife. Silently she looked forward through the rigging, past furled sails, past yards and mast and bowsprit, across the waves toward the unknown north. Occasionally she smiled a little at De Long, rejoicing with him on the surface at least that at last his dream had come true, her eyes shining in her pride in the strength and the love of the man by whom she stood. But what her real feelings were, I, a rough seaman, could only guess, for she said nothing as she clung to the rail, her gaze riveted over the sea to the north into which in a few brief hours were to disappear for ever her husband and her husband's ship.

We stood on a few miles more. The coast line astern became hazy, our escorting fleet of yachts dwindled away to but one. Then a bell jangled harshly in the engine-room below me, the engines stopped. For a few brief minutes the *Jeannette* rolled in the swells while to the shrilling of the silver pipe in the mouth of Jack Cole, bosun, our starboard whale-boat was manned, lowered, and shoved off, carrying toward that lone yacht which now lay to off our quarter, Emma De Long and her husband. There in that small boat, tossing unevenly in the waves a ship-length off, was spoken the last farewell. A brief embrace, a tender kiss, and De Long, balancing himself on the thwarts, handed his wife up over the low side of the yacht.

Another moment, and seated in the sternsheets of the whale-boat, De Long was once more simply the sailor. Sharply his commands drifted across the waves to us:

"Shove off!"

The bowman pushed clear of the yacht.

"Let fall!"

In silence, except for the steady thrash of the oars, our whale-boat came back to us, rounded to under the davits, was hoisted aboard. Again the bell jangled below. Our engines revolved slowly, the *Jeannette* sluggishly gathered speed, the helmsman pointed her west-north-west. Off our quarter, the yacht came about, swiftly picked up headway, and with the two ships on opposite courses, she dropped rapidly astern. For a few minutes, with strained eyes I watched a white handkerchief fluttering across the water at us, then it faded in the distance. The bosun secured the whale-boat for sea, piped down, and without further ceremony, the sea routine on the *Jeannette* commenced. I took a final look at the distant coast and went below to watch the operation of my engines at close range.

CHAPTER V

OF our passage to the Alaskan Peninsula, there is not much to record. It was a shake-down cruise literally enough. The *Jeannette,* between the newly-added weight of her hull reinforcements and the excessive amount of stores and coal aboard, was so grossly overloaded she had hardly two feet freeboard left amidships, and she laboured so heavily in the seas as a consequence that for seagoing qualities, I do her no injustice when I say that as a ship she was little superior to various of Ericsson's iron-clad monitors with which I fell in on blockade duty during the late war. Indeed, had it not been for our masts and spars, our nearly-submerged hull would no doubt have pleased even John Ericsson himself as affording a properly insignificant target for enemy gun-fire.

From our second day out, when the breeze freshened a bit from the north-west and De Long, easing her off a few points to the southward, spread all our canvas to take advantage of it, we were under both sail and steam, but with the seas breaking continuously over our rail and our decks awash most of the time. With our negligible freeboard, we lifted to nothing but took all the seas aboard as they came, rolling heavily and wallowing amongst the waves about as gracefully as a pig in a pen.

In this wise, we discovered a few things, among them the fact that we were burning five tons of coal a day and making only four knots with our engines, which gave us hardly a hundred miles for a day's run. Lieutenant Chipp, an excellent seaman if there ever was one but who had not before been out in the *Jeannette,* was certain he could do as well under sail alone as I was doing under steam, with a consequent saving of our coal, and persuaded the captain to let him try. So below we banked our fires, while the sailors racing through the rigging loosed all our square sails in addition to the fore and aft rig we already had set.

It was interesting to watch Chipp's disillusionment. With all canvas spread up to the fore and main topgallant sails (the *Jeannette* carried nothing above these) Chipp started bravely out on the starboard tack, but in the face of a north-west breeze, he soon found that like most square-riggers, she sailed so poorly by the wind, he had to pay her off and head directly for Hawaii before we began to log even four knots. That was bad enough but worse was to come. Having spent most of the afternoon watch experimenting with the trim of the sails, Chipp finally arrived at a combination to which we logged about four and a half knots, though in the direction of our destination, Unalaska Island, we were making good hardly three. Thus trimmed we ran an hour while the sea-going Chipp in oilskins and boots sloshed over our awash deck from bowsprit to propeller well, his beard dripping water, his eyes constantly aloft, studying the set of every sail from flying-jib to spanker in the hope of improving matters.

In this apparently his inspection gave him no cause for optimism, for after a final shake of the head, he decided to come about and try her on the port tack, to see if by any chance she sailed better there, as is occasionally the case with some ships owing to the unsymmetrical effect of the drag of the screw. Stationing himself amidships, Chipp gave the orders.

Down went the wheel. Then came the final shock. To his great discomfiture, Chipp was wholly unable to bring the lumbering *Jeannette* into the wind and come about! Twice he tried, only to have the ship each time hang "in irons" with yards banging and sails flapping crazily till she fell off again and picked up headway on her old tack. After two failures, De Long tried his hand at it, then Danenhower made an attempt, but in spite of the nautical skill of all her deck officers and the smart seamanship of her crew, the *Jeannette* simply could not be made to tack. For all they could do under sail, the *Jeannette* might still be on that starboard tack headed for Honolulu, if the captain had not finally given up in disgust and roared out to his executive officer in the waist of the ship struggling with the sheets to square away for another attempt.

"Belay that! Leave her to me, Chipp! I'll tack this tub!" and reaching for the bell-pull, he rang the engine-room:

"Full speed ahead!"

Knowing De Long's impetuous nature, I had for some time been suspecting such a result and in the engine-room, I had both coal-heavers and engineers standing by, so I was ready with both boilers and machinery. I yanked open the throttle myself, and our back-acting connecting-rods began to shuttle athwartships. Quickly our shaft came up to fifty revolutions. Above I heard the captain bellow:

"Hard a' lee!"

This time, driven by her screw, the *Jeannette* maintained her headway, came obediently up into the wind, fell off to starboard, and quickly filled away on the port tack. When I poked my head above the machinery hatch coaming to observe results, there was the crestfallen Chipp just outboard of me, busily engaged in securing all on his new course, and I could not resist, a little maliciously, suggesting to him:

"Hey, brother! You want to stay at sea? Well, while you're still young enough to learn, take my advice and study engineering. Sailing ships? In a few years, they'll all be as dead as triremes! Better start now. Let me lend you a good book on boilers!"

But Chipp, still hardly willing to believe that he was beaten, seized a belaying-pin, waved it in my direction, retorted hotly:

"Get below with your greasy machinery and sooty boilers! They're the ruination of any vessel! Sails dead, eh? Unship that damned propeller of yours and I'll tack her!" He jammed his sou'wester viciously down over his ears and ignoring my offer strode forward to check the set of the jibs.

De Long, leaning over the bridge, peering down at me over his dripping glasses, took my gibe at sails more philosophically.

"Well, chief," he observed, "she handles now like nothing I ever sailed in before, but I suppose it's my fault, not hers, she's so low in the water. When we've burned some of this coal and lightened up, perhaps she'll do better." He puffed meditatively at his pipe while he turned to examine the compass. We were hardly within six points of the wind. In dismay De Long muttered: "Heading north! This course will never do if we want to get to Alaska this season. I guess we'll have to douse sail, stick to the engines, and lay her dead into the wind, west-nor'-west for Unalaska, till the breeze shifts anyway." He cupped his hands, shouted after his first officer:

"Mr. Chipp!"

Chipp, just passing the fore shrouds, turned, looked inquisitively up at the bridge.

"Mr. Chipp, furl all sail! We'll proceed under steam alone till further orders!"

For three days we kicked along with unfavourable weather through rain, mist, and head seas. The *Jeannette* laboured, groaned, and with no canvas to steady her, rolled and pitched abominably. Our two scientists, Newcomb and Collins, at the first roll went under with seasickness, and as the weather grew worse their misery passed description, though as is usual in such cases, they got scant comfort from the rest of us. What queer quirk of the sea-going character it is that makes the sailor, ordinarily the most open-hearted and sympathetic of human beings, openly derisive of such sufferings, I know not, but we were no exceptions, and towards Collins in particular, whose puns had occasionally made some of us in the ward-room self-conscious, we were especially barbed in our expressions of mock solicitude.

But if the seasickness of our men of science excited only our mirth, no such merry reaction greeted our discovery that Ah Sam, our Chinese cook, was also similarly indisposed. At first that Ah Sam was seasick was solely a deduction on our part to account for his complete disappearance from the galley, and the fact that for meal after meal we had to make out in the mess-room with only such cold scraps as Charley Tong Sing, the steward, dished out. But when two days went by thus, Ah Sam's whereabouts became a matter of concern to all of us and especially to the doctor. Still, where he had stowed himself, even the bosun could not discover, and to all our inquiries about Ah Sam and to all our complaints about the food, we got from Charley only a shake of the head and in a high sing-song the unvaried reply:

"Ah Sam, he velly sick man now. Cholly Tong Sing, he no feel so good too."

With this unsatisfactory state of affairs in our supply department we had perforce to remain content, until after three days of total eclipse, Ah Sam rose again, one might say, almost from the dead.

I had just come up from the humid engine-room to the main deck, and still bathed in perspiration, had paused to get my lungs full of fresh salt air before diving aft into the shelter of the poop. For a moment, with the wind blowing through my whiskers, I clung to the main shrouds. With both feet braced wide apart on the heaving deck, I stood there cooling off, when from the open passage to the port chart-room, which was the as yet unused workroom for Newcomb's taxidermy, I heard in the doctor's unmistakable Virginian accent:

"Well, I'll be damned! Lend a hand here, Melville!"

I poked my head through the door. Ambler, who I afterwards learned had been tracking down the source of some mysterious groans, had pulled open a locker beneath the chart table, and there, neatly fitted into that confined space, was the lost Ah Sam!

I gasped. Have you ever seen a seasick Chinaman? The combination of Ah Sam's natural yellow complexion with the sickly green pallor induced beneath his skin by *mal de mer,* gave him a ghastly appearance the like of which no ordinary corpse could duplicate. To this weird effect his shrunken body contributed greatly, for he had certainly disgorged everything he had ever assimilated since emigrating from China. (The evidence for this apparently exaggerated statement was such as to convince the most incredulous, but I will not go into details.)

Surgeon Ambler, holding his nose with one hand, grabbed Ah Sam by the pigtail and unceremoniously jerked him forth. Immediately Ah Sam sagged to the deck, his eyes rolling piteously.

Also holding my nose, I seized our cook by one shoulder and dragged him out on deck, where the surgeon gave him a dose of chloroform, which composed him somewhat. Why he had not already died, shut up in that locker, I cannot comprehend, except on the assumption that he grew up in one of those stinkpot factories which Chipp, who was an authority on that country, claimed Chinese pirates maintain. But fearful lest he die yet unless kept out in the air, the captain planked him down at the lee wheel, where under the constant eye of the helmsman, he could not again crawl off to hide in some glory-hole. There, clutching with a death grip at the spokes as the wheel spun beneath his fingers, Ah Sam stayed till next day, a fearful sight with pigtail flying in the breeze and eyes almost popping from their sockets each time a green sea came aboard; and whenever she took a heavy roll, poor Ah Sam's lower jaw sagged open and his tongue cleaved to the roof of his mouth. Such a picture of abject despair and utter anguish as that Chinaman, I never saw.

The fourth day out the weather cleared, moderated somewhat, and the wind shifted to the north-east, so that assisted again by our sails, running on the starboard tack and keeping our desired course, we made a little over five knots for a day's run of one hundred and thirty miles.

As we steamed toward Alaska, we gradually settled down with Chipp labouring continuously to get everything properly stowed. On deck, Ice-Pilot Dunbar, Bosun Cole, and Ice-Quartermaster Nindemann were designated as watch officers, with the seamen under them divided into two watches of four hours each. Below in the black gang, I divided my little force of six men into two watches of six hours each with Machinist Lee and Fireman, 1st Class, Bartlett in charge of the other two men comprising each watch.

For twenty-three days we stood on toward Unalaska Island in the Aleutian chain, a run of about two thousand miles from San Francisco. With better weather, our various landsmen began to show on deck, having acquired sea-legs of a sort. Newcomb, first up, fitted out the port chart-room as his taxidermy shop, spread his tools, and went fishing over the side for albatross almost as soon as he was able to drag himself out of his bunk. With a hook and a line baited with a chunk of salt pork towing astern in the broken water of our wake, he waited patiently but without results while an occasional bird wheeled overhead, till at last, still wan from retching, he turned in, leaving his hook overboard. But Newcomb, whom the doctor and I (chaffing him on his Yankee accent) had nicknamed "Ninkum," was decidedly game. It needed no more than a call from Ambler or me on sighting a new albatross eyeing from aloft that bit of salt pork, of:

"Hey, Ninky, quick! Come and catch your goose!" to bring little Newcomb, aflame with scientific ardour, tumbling up from the poop to man his line hopefully.

At last an albatross measuring some seven feet in wing-spread, which for this ocean is good-sized, swooped down and swallowed the bait, and a bedlam of cries from the anguished bird ensued which attracted the notice of all hands. Then came a battle, in which for a while it seemed debatable whether the albatross, flapping its huge wings frantically at one end of the line, would come inboard, or whether little Newcomb, not yet wholly up to par, tugging on the other end, would go overboard to join it amongst the waves. But Newcomb won at last, landed his bird, promptly skinned it, and prepared it for mounting. And so much has the prosaic power of steam already done to kill the ancient superstitions of the sea, that this albatross, ingloriously hooked, came to its death at the hands of a bird-stuffer without objection or visible foreboding from any mariner aboard.

At last, twenty-three days out of San Francisco, with our bunkers nearly empty from fighting head-winds, and anxious to make port before the coal gave out completely and forced us to rely on our sails alone, we made the Aleutian Islands, only to find them shrouded in thick fog. For Danenhower, our navigator, trouble started immediately. The only chart he had covering that coast was one issued thirty years before by the Imperial Russian Hydrographic Office, and it was quickly apparent that numberless small islands looming up through rifts in the fog were not down on the chart at all, while others he was looking for, were evidently incorrectly located.

Worst of all, Danenhower could not even accurately determine our own position, for when the sun momentarily broke through the clouds, the horizon was obscured in mist, and when the fog lifted enough to show the horizon, the sun was always invisible beneath an overcast sky. For hours on end, Danenhower haunted the bridge, clutching his sextant whenever the horizon showed, poised like a cat before a rat-hole, if the sun peeped out even momentarily, to pounce upon it. But he never got his sight.

Between thick fogs and racing tides, with our little coal pile getting lower and lower, we had a nerve-racking time for two days trying to get through Aqueton Pass into Behring Sea, the *Jeannette* anchored part of the time, under way dead slow the remainder, nosing among the islands, often with only the roar of breakers and the cawing of sea-birds on the rocks to give warning through the mists of the presence of uncharted islets. Finally we slipped safely through, and to the very evident relief of both captain and navigator, on Saturday, August 2, dropped anchor in the harbour of Unalaska.

Naturally, our first concern on entering port was coal. A brief glimpse around the land-locked harbour showed the Alaska Fur Company's steamer *St. Paul,* the schooner *St. George,* owned by the same firm, and the Revenue Cutter *Rush,* but not a sign of our coal-laden schooner, the *Fanny A. Hyde,* nor any report of her having already passed northward on her way to our rendezvous at St. Michael's. So I went ashore with Captain De Long to canvass the local fuel situation. We found eighty tons of coal belonging to the Navy, the remnant of a much larger lot sent north some years before, but so deteriorated by now from long weathering and spontaneous combustion as to be in my opinion nearly worthless. Even so, that being all the coal there was I was investigating the problem of getting it aboard when the commander of the Revenue Cutter requested we leave it for his use, because it would be his sole supply in getting back to San Francisco in the autumn. Naturally we in the *Jeannette,* anxious to get as far north as possible before dipping into what our consort was carrying for us, were not wholly agreeable to waiving our claim as a naval vessel to that coal in favour of the Revenue Service, but this difficulty was soon adjusted by the offer of the Alaska Company's agent to refill our bunkers with some bituminous coal he had on hand, his company to be reimbursed for it in New York by Mr. Bennett. This happy solution settled our fuel question for the moment, so while the natives took over the job of coaling ship, I turned my attention for a few hours to making myself acquainted with the island.

After nearly a month at sea, I found Unalaska pleasant enough, with its green hills surrounding the harbour and its small settlement, comprising mainly the houses of the Fur Company's agents and employees, their warehouses, and last and perhaps most important just at this time, a Greek church. It seems that the steamer *St. Paul,* which we found in the harbour on our arrival Saturday, had just come down two days before from the Pribilof Islands loaded with sealskins, and carrying as passengers from those tiny rocks practically all the bachelor sealers of Pribilof in search of what civilisation (in the form of metropolitan Unalaska) had to offer. All Thursday and Friday there was great excitement here as the native belles paraded before the eager eyes of these none-too-critical prospective bridegrooms. By Saturday, most choices were made and the Greek Catholic Church had a busy day as the couples passed in a continuous procession before the altar and the Russian priest tied the knots. There being no inns or dwellings to accommodate the multitude of honeymooners, the problem was simply enough solved by each newly-wedded couple going directly from the church door for a stroll among the near-by hills. By afternoon, only some few of the sealers, idealists undoubtedly, who were unable to discover among the native women anyone to suit them, were still left wandering disconsolately about the town, peering into every female face, in a queer state of indecision which the smiles of Unalaska's beauties seemed unable to resolve.

Between coaling ship, taking aboard furs for winter clothing, and receiving some six tons of dried fish for dog food, we on the *Jeannette* were kept busy for the three days of our stay, while our nights were enlivened by the most vicious swarms of mosquitoes it has ever been my misfortune to encounter, and all attempts to keep them out of my bunk with the ill-fitting bed-curtains were wholly futile. My bald spot was an especial attraction for them; in desperation I was at last forced to sleep in my uniform cap.

On August 6, we hoisted anchor and got under way, with the whole town on the waterfront to see us off amidst the dipping of colours and a salute from three small guns in front of the Fur

Company's office. I made out plainly enough in the crowd the Russian priest with his immense beard, but De Long and I differed sharply over the presence of any of the brides amongst the throng. So far as I could judge, there were no women there, merely a large crowd of men waving enviously after us as we circled the harbour on our way toward Arctic solitude.

With our usual luck, we bucked a head-wind for all the first day out, but to our great gratification, on the second day the wind shifted to the southward and freshened so that we logged the almost unbelievable day's run of a hundred and seventy-three miles for an hourly average of seven and a quarter knots—for the *Jeannette* almost race-horse speed! But it was too good to last! Next day we had dropped down to a little under six knots, and then the breeze failed us altogether and we finished the last three days of our run to St. Michael's with our useless sails furled, under steam alone at our usual speed of four knots.

CHAPTER VI

THE Kuro-Si-Wo Current, the "black tide" of Japan, somewhat akin to our Gulf Stream, rises in the equatorial oceans south of Asia, flows eastward, is partly deflected northward by the Philippines, and then, impelled by the south-west monsoons, flows at a speed reaching three knots past Japan in a north-easterly direction, a deep blue stream some twelve degrees warmer than the surrounding Pacific Ocean. It was a commonly accepted belief that eastward of Kamchatka, it separated into two branches, one flowing southward along the west coast of North America to temper the coasts of Alaska and British Columbia, while the second branch continued northward through Behring Strait into the Arctic Ocean.

As is well known, for several centuries most of the attempts to reach the North Pole had gone by way of Baffin Bay and Greenland, where without exception they were all blocked by ice. Ours was the first expedition to make the attempt by way of Behring Sea, De Long being willing to test the theory that the warm waters of the Kuro-Si-Wo, flowing northward through the Arctic Ocean, might give a relatively ice-free channel to a high northern latitude, perhaps even to the Pole itself; while if it did not, the shores of Wrangel Land (of which next to nothing was yet known), stretching northward and perhaps even crossing the Pole to reappear in the Atlantic as Greenland as many supposed, would offer a base in which to winter the ship while sledge parties could work north along its coasts toward the Pole.

On these two hypotheses rested mainly our choice of route. With the *Jeannette* in the Behring Sea at last, it remained only to pick up our sledging outfit and put our theories to the test. So for St. Michael's on the mainland of Alaska we headed, where six hundred miles to the northward of Unalaska on the fringe of the Arctic Circle our dogs awaited us and our rendezvous with the *Fanny A. Hyde* was to take place.

The passage took us six days, and many were the discussions round our ward-room mess-table while we steamed on through Behring Sea approaching the real north, as to the correctness of these theories. Especially heated were the arguments with respect to the extent of Wrangel Land whose very existence some polar authorities doubted altogether, since the late Russian Admiral Wrangel (for whom it was named) in spite of a most diligent search, egged on by native reports, never himself was able to find it. As for Kellett and the whaler Long, who afterwards and some years apart claimed to have seen it and even to have coasted its southern shores, they were not everywhere believed.

Aside from these uncertainties, speculation waxed hot over a secondary object of our voyage, to us an unfortunate but unavoidable complication to our task, a search for Professor Nordenskjöld, a Swedish explorer. Attempting that sixteenth-century dream, never yet realised, of the North-east Passage from Europe to the Orient via the Siberian Ocean, he had sailed northward the year before us in the *Vega* from Stockholm to circumnavigate Asia. Nordenskjöld, so it was reported, had successfully reached by the winter time of 1878 Cape Serdze Kamen on the coast of Siberia only a little north of Behring Strait, where almost in sight of his goal, he was frozen in. Since then, except for an unverified rumour from the natives of that occurrence, nothing further had been heard of him or his ship and naturally both in Sweden and in Russia there was considerable anxiety over his fate.

As a consequence, before sailing from San Francisco, we had been ordered by the Secretary of the Navy to search off Cape Serdze Kamen for Nordenskjöld, to assist him if necessary, and only after assurance of his safety, to proceed northward on our own voyage. But we were hopeful that because of the very open summer reported at Unalaska by whalers coming in from the north, Nordenskjöld had been enabled to resume his voyage southward and that we should on our arrival at St. Michael's obtain some definite news of his safe passage through Behring Strait, thereby obviating the necessity of our dissipating what few weeks were left of summer weather in searching the Siberian coasts for him instead of striking directly for the Pole with the *Jeannette* while the weather held.

So one by one, the days rolled by till on August 12 we finally dropped our mud-hook in St. Michael's. After securing my engines, I came on deck to find De Long turning from the unprepossessing collection of native huts and the solitary warehouse which made up the Alaska Company's settlement there, to survey gloomily the empty harbour. Here he had confidently expected to find the *Fanny A. Hyde* waiting with our coal, but no schooner was anywhere visible.

Instead of the schooner, the only boat in sight was a native kyack from which as soon as the anchor dropped, clambered aboard for his mail Mr. Newman, the local agent, who had about given up hope of seeing us this year.

That our schooner had not arrived was evident enough without discussion. But when De Long learned from Newman that they had no tidings whatever of Nordenskjöld, that they had had so far this season no communication with Siberia, and that at St. Michael's they knew even less of Nordenskjöld and his whereabouts than we when we left San Francisco, it was obvious from the droop of the skipper's moustaches that his depression was complete. No schooner, no coal, and now the prospect of having to search Siberia for Nordenskjöld instead of going north!

De Long, as I joined him at the rail to greet Mr. Newman, was polishing his eye-glasses on the edge of his jacket. Meticulously replacing them on his nose as I came up, he sourly scanned the settlement ashore.

"A miserable place, Melville! Look at those dirty huts. Only four white men and not a single white woman here, so the agent says." He turned to the Fur Company agent, added prophetically: "Yet do you know, Mr. Newman, desolate as that collection of huts there is, we may yet look back on it as a kind of earthly paradise?"

Already immersed in his long-delayed mail from home, Newman nodded absent-mindedly. Apparently he was under no illusions about life in the far north.

The captain shrugged his shoulders, philosophically accepted the situation, and after some difficulty in dragging Newman's attention away from his letters from home, we got down to business with the agent, which of course was coal. It developed immediately that St. Michael's had only ten tons of coal, which were badly needed there for the winter. This was hardly a surprise as we had every reason to expect some such condition, but it settled any vague hopes we had that we might coal and proceed before our schooner came. We resigned ourselves to waiting for the *Fanny A. Hyde*.

Next came the matter of our clothing. On that at least was some compensation for our delay. Through Mr. Newman, arrangements were made to send ashore all the furs we had acquired at Unalaska and have the natives (who were experts at it) make them up for us into parkas and other suitable Arctic garments, instead of having each sailor of our crew (who at best had only some rough skill with palm and needle on heavy canvas) attempt with his clumsy fingers to make his own.

With that arranged, the while we waited for our schooner, we settled down to making the best of St. Michael's, all of us, that is, except De Long, who, chafing visibly at the delay, thought up one scheme after another of expediting matters. But each one involved ultimately burning even more coal than waiting there, so finally the baffled skipper retired to his cabin to await as best he could our coal-laden tender.

But even for the seamen, making the best of St. Michael's soon palled and they gave up going ashore. A liberty meant nothing more than wandering round in the mud and the grass, for the village had nothing more to offer a sailor. Even liquor, the final lure of such God-forsaken ports when all else fails, was here wholly absent, its sale being illegal in Alaska Territory. The illegality our seamen knew about, but the absence they refused to believe till a careful search convinced them that the negligible communication of this spot with civilisation made it the one place in the wide world where the laws prohibiting liquor were of necessity observed.

So every other distraction failing, we were thrown back on fishing, the sailor's last resource. Out of curiosity, we set a seine alongside the *Jeannette*. The amount of salmon and flounders we caught opened my eyes—we easily hauled in enough each cast to keep the whole crew in fresh fish every meal, till our men were so sick of the sight of fish that the little salt pork or canned meat served out occasionally from our stores was a welcome change. I see now why these waters are the world's best sealing grounds—they are literally alive with food for the seals, which by the millions swarm over the islands in these shallow seas. The steamer *St. Paul* which we had fallen in with at Unalaska on her way back to America, had her hold packed solid with sealskins, one hundred thousand of them in that vessel alone, a treasure ship indeed!

While the sailors fished, we in the ward-room cast about in various ways for diversion. Newcomb (whom privately the captain was already beginning to regret having brought along, for not only did Newcomb seem never to have grown up but it was now too late to hope that he ever would) went into business for himself. Reverting to the habits of his forbears in far-off Salem, he went ashore with a five-dollar note, purchased from the Alaska Company's store a variety of needles, thread, and similar notions, carted them a mile or two up the coast well out of sight of St. Michael's, set up a "Trading Post," and proceeded to sell his wares to the innocent Indians at just twice what the company store was asking for them.

For this piece of sharp practice at the expense of the natives who were helpfully engaged in making up our fur clothing, gleefully related to the ward-room mess on his return aboard, Newcomb earned the immediate contempt of his fellow New Englander, Dunbar, who burst out:

"You damned Yankee pedlar!" And from that day on, our ice-pilot who himself hailed from the land of the wooden nutmegs and was therefore perhaps touchy of making New England's reputation any worse, refused again to speak to Newcomb, though some of the rest of us including myself, felt with Newcomb that there was at least some humour in the situation.

Tiring of fish and of St. Michael's, I organised a duck-hunting party with Dr. Ambler, Dunbar and Collins for my companions. For a while, I hesitated over including Collins, for by now I had discovered he also had a serious flaw in his character—his sole idea of humour was getting off puns, and so far all the attempts of his shipmates in the ward-room to cure him of it had failed. But as Collins was also our best hand with a shot-gun, I decided to stand the puns for a few hours on the chance of increasing our bag of game and asked him to go.

We purposely took a tent and camped ashore all night to be ready for the ducks at dawn. We got about a dozen (Collins knocked down most of them) but without blinds to work from or decoys to attract our game, it was a tough job and we tramped a long way along the marshy beaches looking for game. During this search we separated, and I with my shot-gun at "ready" was scanning the beach for ducks just below a small bluff, when suddenly there came sliding down its precipitous slope on all fours, face first with hands and feet spread out in the mud in a ludicrous attempt to stop himself, our meteorologist, Collins!

The spectacle was so comical that unthinkingly I roared out to Ambler.

"Look at the old cow there, sliding down the hill!" but I soon enough regretted my outburst for it was evident that Collins, plastered with mud from his mishap and in no humour to see anything funny in his antics, was furious and took my remark as a deep personal insult. So all in all, my hunting party was no great success, and by the time I signalled our cutter to stand in and pick us up, we were all so stiff from sleeping on the hard ground, so throbbing in every muscle from our tramp, and so sullenly did Collins keep eyeing me, that I began to doubt whether a dozen ducks were worth it.

Dr. Ambler, lolling back on the cushions in the stern-sheets of the cutter, homeward bound, apparently took a similar view.

"About once a year of this satisfies me completely, chief." He paused, ruefully massaged his aching calves, then in his careful professional manner continued, "As a doctor, I'm convinced that man's an animal that must take to hard work gradually. No more plunging headlong into it for me! I prescribe a day's absolute rest in our berths for all hands here the minute we hit the ship!"

The doctor, I believe, followed his own prescription, and perhaps Collins and Dunbar did too, but I didn't have time. We had broken

D

a pump-rod on our way to Alaska, temporarily stopping our boiler feed. In that emergency, the spare auxiliary I had installed at Mare Island was immediately cut in on the feed line, saving us from hauling fires and going back to sail alone, but it left us with no reserve pump and it was up to me somehow to provide another rod. Neither Unalaska nor St. Michael's could help me in the least— a machine shop in those primitive trading posts had never even been dreamed of.

With the help of Lee, who was a machinist, and of Bartlett, fireman, first-class, I now set about supplying a new pump-rod from our own resources. While at Mare Island, in view of the uncertainty of repair faculties in the Arctic, like prudent engineers we had acquired for the *Jeannette* quite a set of tools. I won't exactly say we stole them, for after all they merely moved from one spot owned by Uncle Sam to another also under his jurisdiction, but at any rate, in good old Navy fashion during our stay at the Navy Yard everything not nailed down in the machine shop there that appealed to us and that we could carry, somehow moved aboard the *Jeannette,* and now all our recent acquisitions came in handy. I rigged up a long lathe. Out of some square stock once intended by the Navy Yard for forging out chain plates for the *Mohican,* we turned out a very favourable replica in iron of our broken rod, squared off the shoulders for the pistons, cut the threads for the retaining nuts, and long before the schooner showed up in port, had the disabled pump reassembled with the new rod and banging lustily away on the line once more, hammering feed water into our steaming boiler, thus making good my promise to the captain when the old rod broke. This particularly pleased De Long, who I am afraid, like most Line officers, underestimating the resourcefulness of Navy engineers and particularly Scotch ones, had been fearful that we might have to turn back or at least take a long delay while we awaited the arrival, on the *St. Paul's* return trip, or a new rod from the United States.

For six days we waited in St. Michael's, eyes glued to the harbour entrance, undergoing as the captain feelingly expressed it that "hope deferred which maketh the heart sick," when at last on August 18 the *Fanny A. Hyde* showed up, beating her way close-hauled into the harbour. She was a welcome sight not only to our care-worn skipper but to all of us, who long before had completely exhausted in a couple of hours the possibilities of St. Michael's, and in our then state of ignorance, were eager to move on into the even more barren Arctic.

In fact, so eager were we to be on our way that the captain signalled the schooner not to anchor at all but to come alongside us directly, prepared immediately for coaling.

The next three days were busy ones for all hands, lightering coal in bags up from the schooner's holds, dumping it through the deck scuttles into our bunkers, and there trimming it high up under the deck beams to take advantage of every last cubic inch of the *Jeannette's* stowage space. Most of this work of muling the coal around we had to do with our own force, for the schooner with a crew of six men only and being a sailing vessel, with no power machinery of any kind, could assist us but little. Here our deck winch, made of those old steam-launch engines which I had fitted aboard at Mare Island, came in very handy in saving our backs, for with falls rigged from the yard-arms by our energetic Irish bosun, I soon had the niggerheads on that winch whipping the bags of coal up out of the schooner's holds and dropping them down on our decks in grand style.

Needless to say, however, with coal littering our decks and coal-dust everywhere, with state-rooms and cabins tightly sealed up to prevent its infiltration, and with our whole crew as black as nigger minstrels, we carefully abstained from taking aboard any other stores and least of all our furs or dogs from ashore, till coaling was completed and the ship washed down.

At this coaling we laboured steadily until late on the twentieth of August when checking the coal we had already transferred and what was left aboard the schooner, I came to the conclusion that there would still be twenty tons remaining on the *Fanny A. Hyde* for which we could find no stowage, even on our decks, and entering the captain's cabin, I suggested to him that instead of dismissing our escort at St. Michael's as intended, he take a chance and order her to follow us on our next leg, the three-hundred-mile journey across Norton Sound and Behring Strait to St. Lawrence Bay in Siberia, where that last twenty tons of coal she carried, which otherwise would go back to the United States, would just about replenish what we burned on the way over to Asia.

To put it mildly, when I sprang this suggestion on him De Long greeted it with a cheer, but he went me one better.

"That twenty tons she'll certainly carry along for us, chief, but that's not all! What's left in her now, and how long'll it take you to get her down to that last twenty tons?"

"She's got fifty tons still aboard her, captain," I answered. I looked at my watch. It was getting along toward evening already. "But the last thirty tons which we can take aboard from her, will go almighty slow! Trimming it down inside those stifling bunkers to top 'em off for a full due is the devil's own job—it'll take us all day to-morrow certainly!"

De Long, who, downcast over the non-arrival of the schooner, had not cracked a smile for a week, now stroked his long moustaches gleefully.

"Fine, chief! Pass the word to Lieutenant Chipp to belay any more coaling. He's to knock off immediately and start washing down. Here's where we get back one of those lost days, anyway." De Long regarded me with positive cheerfulness. "We'll sail to-morrow! If the *Fanny Hyde's* going to carry twenty tons for us to Siberia, she might as well carry the whole fifty that's still aboard her! So instead of coaling here any more, we'll quit right now, swing ship in the morning to check our compasses, then load furs, stores, and dogs in the afternoon, and sail to-morrow night from this God-forsaken hole! How'll that suit you, chief?"

"Brother, full ahead on that!" I exclaimed. "You'll never get St. Michael's hull down any too soon for me!"

So to the intense relief of the crew, Jack Cole was soon piping down coaling gear. The schooner cast loose, shoved off, and anchored clear, and as darkness fell the hoses were playing everywhere over the *Jeannette's* topsides, washing down, while from every scupper a black stream poured into the clear waters of the bay, as a welcome by-product effectively putting an end to any more fishing in our vicinity.

Our last day at St. Michael's was perhaps our busiest.

In the morning, steaming slowly round the harbour, we swung ship for compass deviations, with Danenhower hunching his burly shoulders constantly over the binnacle while Chipp at the pelorus took bearings of the sun. By noon this essential task was completed and we anchored again, commencing immediately after mess to stow gear and receive stores from ashore.

The display of furs we received, made up now into clothing, of seal, mink, beaver, deer, wolf, Arctic squirrel, and fox, all to be worn by rough seamen, would have caused pangs of jealousy among the ladies on Fifth Avenue, who would have lingered long over each sleek garment, lovingly caressing its velvety softness. But instead of that, disregarded by everyone in our haste, down the hatch shot our furs, our only concern being to get them aboard and weigh anchor.

Following the clothing came aboard assorted cargo—forty Eskimo dogs, five dog sleds, forty sets of dog harness, four dozen pairs of snow-shoes, sixty-nine pairs of sealskin boots, ton after ton of compressed fish for dog food, three small Eskimo skin boats called baideras, and numberless odds and ends; while to top off all, as a personal gift Mr. Newman insisted on presenting to the captain a very handsomely silver-mounted Winchester repeating rifle and eight hundred rounds of ammunition for it.

Last but not least important, came aboard some new members of our crew, two Alaskan Indians from St. Michael's. This pair, Alexey and Aneguin, carefully selected on the recommendation of the entire white population of St. Michael's (all four of them), were after a lengthy pow-wow over terms with the head-man of the native village shipped as hunters and dog-drivers. Alexey, as senior hunter, was to be paid twenty dollars a month; Aneguin, his assistant, as a hunter's mate (to put it in nautical parlance) was to receive fifteen; and each was to draw from the company store an outfit worth fifty dollars to start with and on discharge to receive a Winchester rifle and 1,000 cartridges. To the wife of Alexey and to the mother of Aneguin, thus deprived of their support, were to be issued at the *Jeannette's* expense from the Alaska Company's store, provisions to the value of five dollars each monthly until their men should finally be returned to St. Michael's.

These terms being finally settled to the satisfaction of all, Alexey and Aneguin reported aboard at 5 p.m., both for the first time in their lives dressed in "store clothes" which they had just drawn from Mr. Newman's stock, and proud as peacocks in shiny black Russian hats, topped with flaming red bands. Alexey (who to the best of my knowledge, aside from our captain, was the only married man aboard) was accompanied by his Indian wife, a small, shy, pretty woman in furs oddly contrasting with her husband's stiffly worn civilised raiment, and by his little boy. Tightly holding each other's hands, this tiny Alaskan group drifted wonderingly over the ship, children all in their open-mouthed curiosity; while Aneguin, accompanied by his chief and a delegation of natives come to see him off, was just as naïve in exclaiming over everything he saw, and the excitement of all reached a high pitch when Captain De Long presented to Alexey's shrinking little wife a china cup and saucer with "U.S.N." in gold on it, and to her little boy, a harmonica.

As evening drew on and the hour for departure approached, Alexey and his wife, seated on a sea-chest on the poop, clung silently to each other, till at the hoarse call of the bosun, "All visitors ashore!" accompanied by significant gestures toward the rail, they parted affectionately—and for ever.

For a few minutes there was a grand scramble of Indians over our bulwarks into native boats. Then to the rattling of the chain links in the hawse-pipe, our cable came slowly in and with a blast of our whistle in salute, we got under way for St. Lawrence Bay, on the Siberian side of Behring Sea.

CHAPTER VII

THROUGH a light breeze and a smooth sea we steamed out in the darkness. The *Fanny A. Hyde,* ordered to follow us at dawn, we expected to reach port in Siberia even ahead of our own arrival since she was now very light while we, heavily laden once more, were nearly awash.

A new note in sea-going came into our lives upon departing from St. Michael's—our forty dogs. They quickly proved to be the damnedest nuisances ever seen aboard ship, roaming the deck in carefree fashion, snarling and fighting among themselves every five minutes, and unless one was armed with a belaying-pin in each hand, it was nearly suicide to enter a pack of the howling brutes to stop them. They fought for pure enjoyment so it seemed to me, immune almost from any harm, for their fur was so thick and tangled, they got nothing but mouthfuls of hair from snapping at each other. In spite of fairly continual fighting, we got the ship along, for after all my engines drove her on, but how we should ever fare under sail alone I wondered, unless each seaman soon got the knack of disregarding half a dozen pseudo-wolves leaping at him each time he rushed to ease a sheet or to belay a halliard or a brace. Meanwhile we let the dogs severely alone, it being the duty of Aneguin and Alexey to feed and water them, and apparently also to beat them well so that their fighting was not one continuous performance.

We had expected to make the three hundred miles across Behring Sea to St. Lawrence Bay in Siberia in two and a half days, but we did not. Our second day out, the wind freshened, there was a decided swell from the northward, and all in all the weather had a very unsettled look. With most of our sail set, we logged five knots during the morning, but as the seas picked up, the ship began to pitch heavily, and in the early afternoon a green sea came aboard that carried away both our forward water-closets, fortunately empty at the time. At this mishap, we furled most of our canvas and slowed the engines to thirty revolutions, greatly easing the motion.

But as the night drew on it became evident we were in for it. The ocean hereabouts is so shallow that an ugly sea quickly kicks up under even a fresh breeze, and we were soon up against a full gale, not a pleasing prospect for a grossly overloaded ship. There was nothing for it, however, except to heave to, head to the wind, and ride it out under storm-sails only. Accordingly, in the first dog-watch, I banked fires, stopped the engines, and the *Jeannette* lay to on the starboard tack under the scantiest canvas we dared carry—stormsail, fore and aft sail, and spanker only, all reefed down to the last row of reef-points.

For thirty hours while the gale howled, we rode it out thus, the overloaded *Jeannette* groaning and creaking, submerged half the time, with confused seas coming aboard in all directions. Every hatch on deck was tightly battened down; otherwise solid water, often standing two and three feet deep on our decks, would have quickly poured below to destroy our slight buoyancy and sink us like a rock.

But even so, in spite of lying to, we took a terrific battering, and time after time as we plunged into a green sea, it seemed beyond belief that our overloaded hull should still remain afloat.

In the middle of this storm, worn from a night of watchfulness, with Chipp on the bridge temporarily as his relief, Captain De Long sat dozing in his cabin chair, not daring even to crawl into his bunk lest he lose a second in responding to any call. Suddenly a solid sea came over the side, with a wild roar broke on board, and in a rushing wall of water carrying all before it, hit the poop bulkhead, smashed in the windows to the captain's state-room, and in an instant flooded the room. Our startled skipper coming out of his doze found himself swimming for his life in his own cabin, all his belongings afloat in a tangle about him!

For the first hour of that gale, the howling of our forty Eskimo dogs was a fair rival to the howling of the wind through the rigging, but as the waves began to break aboard, the poor dogs, half-drowned, quieted to a piteous whimper, and with their tails between their legs, sought shelter from the rushing seas in the lee of the galley, the bulwarks, the hatch coamings even—anything that would save them from the impact of those swirling waves. For once there was no fighting, each dog being solely absorbed in keeping his nose above water, and when possible on that heaving deck, in keeping his claws dug into the planking to save himself from being flung headlong into the lee scuppers

But the gale finally blew itself out, and thankfully spreading our reefed canvas, we arrived four days out of St. Michael's in lonely St. Lawrence Bay, to find the little *Jeannette*, a tiny symbol of civilisation, dwarfed in that vast solitude by snow-capped mountains rising precipitously from the water, a magnificent spectacle of nature in her grandest mood.

But our isolation was broken soon enough by two large baideras which pushed out to meet us, crowded with natives who without leave clambered over our rails, eagerly offering in broken English to engage themselves as whalers, which naturally enough they assumed was the purpose of our cruise.

But we welcomed them gladly enough for another reason. What did they know of Nordenskjöld?

From their chief, a tall, brawny fellow calling himself "George," after much cross-examination De Long elicited the information that a steamer, smaller even than the *Jeannette*, had been there apparently three months before, and that during the previous winter he, Chief George, had on a journey across East Cape to Koliutchin Bay on the north coast of Siberia, seen the same ship frozen in the ice there. This seemed to check with our last news on Nordenskjöld's *Vega*. If indeed she had reached St. Lawrence Bay and passed south, she was, of course, safe now and we need no longer concern ourselves. But was it really the *Vega?*

Patiently, like a skilled lawyer examining an ignorant witness, De Long worked on George to find that out. Who was the *Vega's* captain? An old man with a white beard who spoke no English. Who, then, had George conversed with? Another officer, a Russian, who spoke their tongue, the Tchuchee dialect, like a native. Who was he? On this point, George, uncertain over nearly everything else, was absolutely positive, and answered proudly:

"He name Horpish."

But to De Long's great disappointment, on consulting the muster roll of the *Vega* with which we had been furnished, no "Horpish" appeared thereon. Again and again, Chipp, De Long, and I pored over that list of the Swedish, Danish, and Russian names of the men and officers accompanying Nordenskjöld, while George, leaning over our shoulders, repeated over and over, "Horpish, he Horpish," obviously disgusted at our inability to understand our own language.

Finally De Long put his finger on the answer. There, a few lines down from Nordenskjöld on that list was the man we were looking for—

"Lieutenant Nordquist, Imperial Russian Navy."

I pronounced it a few times—Nordquist, Horpish—yes, it must be he. Phonetically in Tchuchee that was a good match for Nordquist.

And this was all we learned. The steamer, whatever her name, had stayed only one day, then departed to the southward, loaded, according to George, with "plenty coals."

With some bread and canned meat in return for this sketchy information, we eased George and his followers, greatly disappointed at not being signed on as whalers, over the side before we lost anything. For while these Tchuchees appeared dirty, lazy, and utterly worthless, their unusual size made them potentially dangerous enemies when in force, and we posted an armed watch on deck as a precaution.

Our schooner arrived soon after we did, and we finished hoisting out of her all the coal down to the last lump, ending up with 132 tons stowed in our bunkers, which was their total capacity, and with 28 tons more as a deck-load, giving us a total of 160 tons with which to start into the Arctic, nearly twice the amount of coal the *Jeannette* was originally designed to carry.

On August 27th we finished coaling and steamed out towing the schooner astern of us, for it seemed unsafe with her little crew of only six to leave her to get under way in desolate St. Lawrence Bay amidst that ugly-looking crowd of brawny Tchuchees, all experts at handling harpoons and looking none too scrupulous over what they chose to hurl them at.

Once clear of the harbour, we headed north for Cape East, while the lightened *Fanny A. Hyde,* carrying now as cargo only the last mail we ever sent back, spread her sails and with (for her but not for us) a fair wind was soon hull down to the southward, our last link with home finally severed.

Steaming steadily into a strong head wind, we stood on through Behring Strait, and during the night passed between East Cape and the Diomede Islands, three barren rocks jutting from the sea, forming stepping-stones almost between the continents of Asia and America, over which may very well have passed ages ago that immigration at the time of the dispersion of the human race which brought man first to North and then to South America.

But this human migration of former ages, even if so, interested us little in comparison with what the migratory waters of the ocean might be doing now. At the captain's orders, Collins prepared a set of thermometers and dropped them overboard strung out on a line. If the Kuro-Si-Wo Current actually flowed northward into the Arctic Sea as we hoped, through this narrow funnel it must pass, and as we steamed slowly northward through the strait, Collins periodically read the thermometers to get the temperatures at varying depths, while Newcomb tended a dredge towed astern to obtain samples of the marine life at the bottom.

To our keen chagrin, the most that could be said for the results of our observations was that they were neutral—they proved or disproved nothing. The water was about the same in temperature from top to bottom and did not differ appreciably from the temperature of the air, a result which certainly did not indicate the presence of any marked warm current thereabouts. But then on the other hand, as we passed through, the fresh breeze we encountered from the north-west, blowing down through the strait, might well on that day have upset or even reversed the normal flow of water in a channel only twenty-eight fathoms deep. The thermometers proved nothing. How about the dredge? Eagerly we awaited a report from Newcomb with respect to his examination of its contents. Were the specimens in any degree symptomatic of the tropical waters of the Kuro-Si-Wo Current?

But there also we got scant comfort. The catch in the dredge was nondescript, and no deductions could safely be drawn. If the Kuro-Si-Wo Current on which we were banking so heavily for the success of our expedition flowed into the polar seas, at least we found no evidence of it.

As the day dawned with the empty horizon widening out before us to the north, we found ourselves at last in the Arctic Ocean, our gateway to the Pole. We stood to the north-west with somewhat overcast skies, coasting along the northern shores of Siberia before striking off for Wrangel Land, our thorough-going captain determined to steam a little out of his way to make one more stop at Cape Serdze Kamen to check Chief George's story of the *Vega's* actually having been there and left.

In the late afternoon, we made out a headland on our port hand, which the vigilant Danenhower, fortunately able just then to catch a sight of the sun to establish our position, pronounced as the desired cape. We stood in and anchored, but with steam up ready for getting under way instantly, since we were on a lee shore with none too good a holding ground.

After supper, the starboard whale-boat was cleared away and lowered, and the captain, backed up by Chipp, Dunbar, Collins and Alexey, started in for a collection of native houses on the beach to investigate about the *Vega*. But when they approached the shore, they found such a heavy sea rolling over a rapidly moving fringe of pack ice that the whale-boat's efforts to get through were wholly ineffectual and after half an hour were abandoned when it was observed that the natives were getting ready to come out themselves in a skin boat.

Led by their chief in a bright red tunic, these latter, better acquainted with the coast than we, managed to make passage through the ice into the open water off shore, where they followed

the whale-boat back to our ship and then over the side into the
Jeannette's cabin.

But there, in spite of Alexey's best efforts with native dialects on
the chief, who had all the dignity of a king, we were unable to
make our questions about Nordenskjöld understood, or get anything understandable out of him except the one word:

"Schnapps?"

But to this De Long shook his head. He was too well acquainted
with the results to pass fire-water out to natives. So with this we
were stalled till the natives pushed forth a decrepit old squaw, who
once an inhabitant of Kings Island, apparently had recognised
Alexey's dialect. But even on her, neither the name of Nordenskjöld
nor of the *Vega* made any impression till Alexey, remembering Chief
George's reference to the Russian officer who spoke Tchuchee, mentioned Lieutenant Nordquist.

Immediately the squaw became all animation and her face lighted
with understanding.

"Horpish?" She nodded her head vigorously and from then on,
all was plain sailing. Alexey had difficulty in getting in a word sideways, "Horpish" had made such an impression. "Horpish's" ship
had been there a day, coming from a little farther west in Koliutchin
Bay, where she had wintered, a fact very soon verified by Chipp,
who going ashore to visit that spot, came back with a miscellaneous
collection of articles—Swedish coins, buttons from Russian, Danish,
and Swedish uniforms, and prized most of all by the natives and
hardest to get from them, a number of empty tin-cans!

This evidence settled conclusively our search for Nordenskjöld
and the *Vega*. Undoubtedly she had been at Cape Serdze Kamen
as rumoured; she had passed undamaged through Behring Strait,
and must now be on her way southward through the Pacific to
Japan, if indeed she had not already arrived there. We need no
longer concern ourselves over Nordenskjöld and his crew, their
safety was assured.

So on the last day of August, we hoisted in our whale-boat and
stood out to the northward. It being Sunday, at 2 p.m. we rigged
ship for church in the cabin, and our captain, a devoutly religious

man, held Divine Service, attended by all the crew save those on watch.

With heads bowed, we stood while our ship steamed away from that bleak coast, our anchor hoisted in for the last time, with our thoughts divided between Nordenskjöld safely homeward bound and ourselves headed at last into the unknown polar seas.

As Collins rolled the notes solemnly out from the little organ and the rough voices of our sailors echoed the words, never before had I seen men at sea so deeply stirred by that heartfelt appeal:

"Oh, hear us when we cry to Thee,
For those in peril on the sea!"

CHAPTER VIII

FOR the first time, ice now began to be a factor in our cruise. We had noted a little along the Siberian shore churned by the surf when the whale-boat attempted a landing off Serdze Kamen, but now as we stood away from the coast, pack ice to the westward making out from Koliutchin Bay bothered the ship noticeably, with loose ice in large chunks bobbing about in the waves, necessitating constant conning by the officer of the watch to avoid trouble. Finally at 10 p.m., with ice growing heavier, while our course to Wrangel Land lay N.-W. by N., the captain changed course to N.-E. for a few hours to take her out of it, and then having come to open water, back to N.-W., on which course under sail and steam we stood on through the night and all next day with beautifully clear cold weather attending us.

About a hundred miles to the southward of where Wrangel Land should be, we made out the ice-pack once more, extending this time from dead ahead uninterruptedly around to the westward as far as eye could see. Confronted thus by the solid pack across our path, there was nothing for it but to head the *Jeannette* off to the eastward, away from our objective, skirting as closely as we dared that pack on our port side, solid ice now seven feet thick!

Meantime a fine south-east breeze sprang up, and to this we made all sail, heading north-east with wind abeam and the ice dead to leeward, while from the crow's-nest, grizzled old Dunbar, our ex-whaler ice-pilot, closely scanned the pack for any lead of open water through it going northward, but he found not the slightest sign of one.

On that course we were constantly increasing our distance from Wrangel Land instead of diminishing it, so De Long after morosely regarding for some time the fine wake which our six-knot speed was churning up in the icy water astern, finally ordered me when darkness fell to stop the engines, bank fires, and save the coal, letting her go under sail alone for the night.

Late in the first watch then, the engines were secured, the fires heavily banked in the boilers to burn as little coal as possible, and stocky Bartlett, fireman in charge of the watch, instructed to keep them so. With all secured below I came up on deck, for a few minutes before turning in looking off to leeward across the black water at the vague loom of that solid ice-pack fringing the near horizon.

Eight bells struck, the watch was changed, the men relieved tumbling below to the forecastle with great alacrity, for in spite of the south-east breeze, there was a sharp chill in the cutting wind as the *Jeannette,* with all sails drawing, plunged ahead at full speed. Deeply laden and well heeled over by a stiff beam wind, we were running with the lee scuppers awash, and the cold sea threatened momentarily to flood over our low bulwark. What with the icy water and the chilly air, the contrast with the warmth of the boiler-room I had just left was too much for me. With a final glance over-head at our straining cordage and taut canvas and a wave to Dunbar who, with dripping whiskers dimly visible in the binnacle light on the bridge above me, had just taken over the watch on deck, I ducked aft into the poop and wearily slid into my bunk.

On the starboard tack with the wind freshening, the *Jeannette* stood on through the night. One bell struck. In the perfunctory routine drone of the sea, the look-out reported the running lights burning bright and the report was gravely acknowledged by Mr. Dunbar, though we might just as well have saved our lamp oil, for what ship was there besides ourselves in that vast polar solitude to whom those lights, steadily burning in the darkness, might mean anything in the way of warning?

Nevertheless, we were under way. Habit and the law of the sea are strong, so on deck the incongruity of the reports struck no one. Hans Erichsen, a huge Dane posted in the bow as look-out, turned his eyes lazily from the gleaming lights in the rigging toward the bowsprit once more, gradually accustoming them again to the darkness ahead.

And then hoarse and loud, nothing perfunctory this time about the call, came Erichsen's cry.

"Ice ho! Dead ahead and on the weather bow!"

On the silent *Jeannette,* that cry, cutting through the whistling of the wind and the creaking of the rigging, echoed aft in the poop to bring up in the twinkling of an eye, tumbling half clad out of their bunks, Captain De Long, Mr. Chipp, and all the other officers.

"Hard alee!" roared Dunbar to the helmsman, desperately endeavouring to bring her into the wind to avoid a collision, for with ice alee, ahead, on the weather bow, there was no way out except to tack.

But the *Jeannette,* heavily laden and with a trim by the stern as she then was, had never successfully come about except with the help of her engines. And now the fires were banked! But she must tack or crash!

"All hands!"

Through the darkness echoed the rush of feet tumbling up from the forecastle, racing to man sheets and braces, the shrill piping of the bosun, hoarse orders, then a bedlam of curses and the howling of dogs as all over the deck, men and animals collided in the night.

In response to her hard over rudder, the *Jeannette's* bow swung slowly to starboard while from ahead, plainly audible now on our deck, came the roar of the waves breaking high on the solid pack.

Would she answer her helm and tack?

Breathlessly we waited while with jibs and headsails eased and spanker hauled flat aft, the *Jeannette* rounded sluggishly toward the wind and the open sea and away from that terrible ice.

Then she stopped swinging, hung "in irons." With our useless sails flapping wildly and no steam to save us, helplessly we watched with eyes straining through the darkness as the *Jeannette* drove broadside to leeward, straight for the ice-pack!

E

CHAPTER IX

WE struck with a shivering crash that shook the *Jeannette* from keel to main truck, and hung there with yards banging violently. Lucky for us now, that nineteen-inch thickness of heavily reinforced side and the stout backing of those new trusses below— that impact would have stove in the side of any ordinary vessel!

But though we had survived that first smashing blow, we were in grave danger. Important with sails and rudder to claw off that ice-bank, we lay there in a heavy seaway, rolling and grinding against the jagged shelf on which the wind was pushing us.

That put it up to the black gang. I rushed below into the fire-room.

"Bartlett!" I yelled. "Wide open on your dampers! Accelerate that draught!"

"Sharvell! Iversen!" I sang out sharply to my two coal-heavers. "Lively with the slice-bars! Cut those banked fires to pieces! Get 'em blazing!"

For thirty anxious minutes we fought before our two Scotch boilers with slice-bars, rakes, and shovels to raise steam, while through our solid sides as we toiled below the water-line, we heard the groaning and the crunching of the ice digging into our planking and from above the slapping of the sails, the howling of the dogs, and the kicks and curses of the seamen still struggling futilely to get the ship to claw off to windward.

At last with fires roaring, the needle of our pressure gauge started to climb toward the popping-point; I reported we were ready with the engines.

De Long doused all sail; under steam alone with our helm hard aport and propeller turning over at half speed, we swung our bow at last to starboard into the wind and slowly eased away from the pack, decidedly thankful to get clear with no more damage than a terrible gouging of our stout elm planking. And under steam alone

54

for the rest of the night we stood on dead slow nearly to windward between east and south-east, keeping that ice-pack a respectable distance on our port hand till dawn came and with it, a fog!

For the next few watches, we played tag with the ice-fields, standing off when the fog came in, standing in when the fog lifted, searching for an open lead to the northward. At one time during this period, the fog thinned to show to our intense astonishment, off to the south-east a bark under full sail, a whaler undoubtedly, standing wisely enough to the southward away from the ice, but so far off, anxious as we were not to lose any northing while we sought an open lead, we never ran down and spoke him.

Soon, a little regretfully, we lost him in the fog, the last vessel we ever saw, homeward bound no doubt and a missed opportunity for us to send a farewell message home before we entered the ice-pack around Wrangel Land.

Finally with nothing but ice in sight except to the south-east, De Long decided to try a likely-looking lead opening to leeward, toward the north-west. So with the captain in the crow's-nest and the ice-pilot perched on the topsail yard, we entered the lead, Lieutenant Chipp on the bridge conning the ship as directed from aloft. Cautiously we proceeded in a general north-westerly direction up that none too wide lead of water with broken ice-fields fairly close aboard us now on both sides, for some seven hours till late afternoon, when simultaneously the lead suddenly narrowed and the fog thickened so much that we stopped, banked fires, and put out an ice-anchor to a near-by floe.

Chilled, cramped, and dead-tired from his long day in the crow's-nest, De Long laid down from aloft and promptly crawled into his bunk, while the fog continuing, we lay to our ice-anchor till next day.

For the first time on our cruise, the temperature that night dropped below freezing, with the odd result that by morning between the fog and the freezing weather, our rigging was a mass of shimmering snow and frost, magically turning the *Jeannette* into a fairy ship, a lovely sight with her every stay and shroud shining and sparkling in the early dawn, and the running rigging a swaying crystal web of jewels glistening against the sky.

But as the fog still hung on, and we consequently could not move, I am afraid our captain, more interested in progress northward than in beauty, gave scant heed, and it was left to Ambler and me, being early on deck, really to drink in the soul-satisfying loveliness of that scene.

Some new ice, a thin film only, made around the ship during the night, the weather being calm and the surface of our lead therefore undisturbed and free to freeze, but it was insignificant in thickness compared to the pack-ice surrounding us, which seemed everywhere to be at least seven feet in depth, of which thickness some two feet were above water and the rest below, with some hummocks here and there pushed up above the smooth pack to a height of six feet perhaps.

By afternoon, the fog cleared enough for us to haul in our ice-anchor, spread fires and get under way along our lead, with running now in a north-easterly direction we followed for two hours, poking and ramming our way between drifting floes. Then to our delighted surprise, we emerged into the open sea again, open, that is, between east and north only, with ice filling the horizon in all other directions.

With some sea-room to work in, we speeded the engines and headed north, where we soon passed a drifting tree, torn up by the roots, an odd bit of flotsam to encounter in those waters, but which as it must have come from the south encouraged us since it lent some weight, however slight, to the Japanese Current theory about which we were beginning to entertain serious doubts. But we had little time to speculate on this, for soon from the look-out came the cry:

"Land ho!"

Sure enough, bearing north-west, apparently forty miles off and much distorted by mirage, was land which from our position and its bearings we judged to be Herald Island. This island I must hasten to explain was so named, not after the *New York Herald* whose owner, Mr. Bennett, was financing our expedition, but after H.M.S. *Herald,* whose captain, Kellett, had discovered and landed on that island thirty years before, in 1849.

Immediately from alow and aloft all hands were scanning the island, through binoculars, through telescopes, and with the naked eye. There was much animated discussion among us as to its distance, but regardless of that we could do nothing to close on it, for the ice-field lay between. So as night fell, we merely steamed in circles at dead slow speed, just clear of the pack.

Day broke fine and crisp with a light northerly breeze off the ice. Picking the most promising lead toward Herald Island, we pushed the *Jeannette* into it, and for two hours amidst drifting floes we made our way with no great trouble, when, to our dismay, we began to meet new ice in the lead, from one to two inches in thickness. For another two hours, we pushed along through this, our steel-clad stem easily breaking a path through which we drove our hull, with the thin ice scratching and gouging our elm doubling, when we came at last smack up against the thickest pack we had yet seen, some ten to fifteen feet of solid ice. This, needless to say, brought us up short. Since we could do nothing else, we ran out our ice-anchor to the floe ahead, while we waited hopefully for some shift in the pack to make us a new opening.

With clearer weather, several times during the morning as we lay in the ice, we made out distinctly not only Herald Island, but other land beyond, above, and also to the south-west of it, which from everything we had been told, should be part of that Wrangel Land on which we were banking so much to afford us a base for our sledging operations toward the Pole. Consequently we searched the distant outlines of this continent with far greater interest than we had bestowed on the nearer profile of Herald Island, but to no conclusion. Danenhower, Chipp, and De Long, all experienced seamen, strained their eyes through glasses, scanning what could be seen of the coast of Wrangel Land, but so far even from agreeing on its remoteness, looking across ice instead of water so upset their habits of judging, that their estimates of its distance varied all the way from forty to one hundred miles, while De Long even doubted whether what he saw beyond Herald Island was land at all but simply a mirage. Being only an engineer, I took no part in these discussions, more concerned myself in staring at the unyielding edges of the near-by floes and wondering, if our navigation for the next few weeks was to consist mostly of traversing leads filled with such floating ice cakes, how long we could hope to go before an ice-floe sucked in under our counter knocked off a propeller blade, and how long a time would elapse before our four spare blades were all used up. But there was no great occasion for such worry on my part. Not till afternoon could we move at all, and then only for a couple of hours, when once more we were brought up by solid ice ahead and with banked fires again anchored to a floe, called it a day, and laid below for supper.

Supper was an unusually sombre meal. Such an early-season encounter with the ice-fields and at so low a latitude, was a sad blow to our hopes of exploration. De Long, at the head of the table, served out silently, as Tong Sing placed the dishes before him; I, on his left, carved the mutton and aided him at serving—to Chipp first, then to the others on both sides of the table down to Danenhower, who as mess treasurer sat at the foot of the table opposite the captain. Potatoes, stewed dried apples, bread, butter, and tea made up the rest of our unpretentious meal, the simplicity of which perhaps still further emphasised our situation and put a damper on any conversation. Only the shuffling of the Chinese steward's feet on the deck as he padded round the little ward-room with the plates broke the quiet.

De Long, brooding over the ship's situation, was gradually struck by the absence of conversation and its implications. More I think to make conversation than in the hope of gaining any information, he picked out the ice-pilot on my left, sawing earnestly away at his mutton and asked him:

"Well, Mr. Dunbar, do you think we'll get through this lead to Herald Island?"

Dunbar, absorbed like the rest of us in his thoughts, surprised me by the speed, so unusual for him, with which without even looking up he snapped out his reply.

"No, cap'n, we won't!" Then more slowly as he turned his grizzled face toward the head of the table, he added vehemently, "And what's more, while God's giving us the chance, I'd wind her in that little water hole astern of us and head out of this ice back to open water before the bottom drops out of the thermometer and we're frozen in here for a full due!"

Astonished by the heat of this unexpected reply, De Long looked from the old whaler, who in truth had hurled a lance into the very heart of each man's thoughts, to the rest of us, all suddenly straightened up by the thrust.

"And why, Mr. Dunbar?" in spite of a pronounced flush he asked mildly. "Where can we do better, may I ask?"

"Farther east, off Prince Patrick's Land, to the north'ard of the coast of North America," replied Dunbar shortly. "A whaler'll stay in open water farther north'n this over on the Alaska side most any time; the current sets that way toward Greenland, not this side toward Siberia."

De Long calmly shook his head.

"No use, pilot; we're not whaling and we'll not go east. That would take us away from Wrangel Land, and sledging north along the coasts of Wrangel Land's our only hope for working into the real north from Behring Strait. No, we can't do it. We'll have to take our chances here."

Dunbar, his suggestion overruled, made no reply, masking his disappointment by hunching a little lower over his plate and hacking away once more at the chunk of mutton before him. And as suddenly as it had flared up, all conversation ceased.

September 6 dawned, for us on the *Jeannette* a day to which we often looked back with mingled feelings. During the night our water lead froze up behind us. In the morning, as far as the eye could see in every direction now was only ice—no water, no open leads anywhere. A fog hung over the sea, blotting out Herald Island, but a light northerly wind gave some promise of clearing the atmosphere later on.

We gathered at breakfast in the cabin, a sombre group. Under way for a week since leaving Cape Serdze Kamen, we had made but 240 miles to the north, to reach only lat. 71° 30′ N., a point easily to be exceeded by any vessel all year round in the Atlantic. But here we were, completely surrounded by ice. Was this the exceptionally open Arctic summer, so free of ice, that in Unalaska we had been informed awaited us?

Danenhower, loquacious as always, broke the silence, observing to no one in particular:

"This damned coffee's even worse than usual, all water and no coffee beans. Ah Sam's had time enough to learn by now. Can't anyone persuade that Chink to put *some* coffee in the pot? What's he saving it for?"

"Maybe the sight of all that ice discourages him," observed Ambler. "Perhaps he thinks we're in for a long hard winter and he's got to save. I reckon he's right too, for that ice-pack sure looks to me as if it never has broken up and turned to water yet."

"Right, surgeon." Captain De Long at the head of the table, busily engaged in ladling out a dish of hominy, looked up at Ambler and nodded pessimistically:

"And what's worse for us, it looks to me as if it never will, unless someone whistles up a heavy gale to break up the pack."

Chipp, uncomplainingly engaged in drinking down his portion of the insipid coffee, took objection at this.

"Don't try that, captain! In any gale that'd break up *this* pack the pack'd break the *Jeannette* up in the process. No, let Nature take her course melting that ice; it may be slower but it's safer."

"Come down to earth!" broke in Danenhower. "Let's leave the pack a minute; it'll be there for a while yet. I was talking about coffee. Hasn't anybody in this mess got influence enough to get Ah Sam to pack a little coffee in the pot for all this water to work on?"

"Well," grinned Collins, seeing a chance to slip in a pun, "you're the navigator, Dan. Why don't you try shooting that Celestial's equator? That ought to stir him up."

Collins, chuckling happily, glanced round for approval.

Danenhower twisted his broad shoulders in his chair, directed a blank stare at Collins.

"Huh? If that's another one of your puns, Mr. Meteorologist, what's the point?'"

Collins stopped laughing and looked pained.

"Don't you see it, Dan?" he asked. "Why, that one's rich! Celestial, equator, and you're a navigator. Now, do you get it?"

Danenhower, determined with the rest of us to squelch Collins's puns, looking as innocent of understanding as before, replied flatly:

"No! I'm too dumb, I guess. Where's the point?"

"Why, Ah Sam's a Chinaman, isn't he?"

"If he's not, then I'm one," agreed Danenhower. "So far I'm with you."

"Well, all Chinamen are sons of Heaven, aren't they? So that makes him a Celestial. See? And you're a navigator so you shoot the stars; they're celestial too. And anybody's stomach's his equator, isn't it? You see, it all hangs together fine. Now do you get it?" inquired Collins anxiously.

"I'm damned if I see any connection in all this rigmarole of yours with my attempts at getting better coffee," muttered Danenhower. "Does anybody?" He looked round.

Solemnly, first De Long, then Chipp, Ambler and I all shook our heads, gazing blandly at Collins for further elucidation as to what the joke might be.

Collins looked from one to another of us, then in disgust burst out:

"The farther all of you get from San Francisco, the weaker grow your intellects!' He leaned back sulkily. "By the time we get to the Pole, you won't know your own names. Why, that one's good! They'd see it in New York right off. I've half a mind to try it out on that Indian, Alexey. I'll bet even he sees it!"

"Why don't you try it on Ah Sam instead, then?" queried Danenhower, rising. "If our cook sees it, there's hope. Maybe next you can make him see why he ought to put some coffee berries in the pot when he makes coffee, and that'll be something even my thick skull can understand!" He jerked on his pea-coat, lifted his bulky form from his chair, and strode to the door. "I'm going on deck. I'm too dumb, I guess, to see the points of Collins's puns. But maybe if I'm not too blind yet, I can see the ice, anyway."

With a wink at Ambler, our navigator vanished. It seemed to be working; perhaps we might yet cure Collins of his continuous stream of puns, for most of them were atrocious, and anyway, having now had a chance to get acquainted at close range with punning, I heartily agreed with whoever it was, Samuel Johnson I think, in averring that a pun was the lowest form of wit. With us the case was serious—here with the long Arctic night approaching, locking us within the narrow confines of our vessel, we were shipmates with a punster and no escape except to break him of it!

I rose also and went out on deck, the while Collins turned his attention to Dunbar, trying to get him, who also knew something of navigation, to admit that he at least saw the point in the meteorologist's play on words, but I am afraid he picked the wrong person, for Dunbar's grim visage remained wholly unresponsive.

Out on deck, clad in a heavy pea-coat with a sealskin cap jammed tightly down over my bald spot, for the temperature was down to 26° F., I looked around. A distant view was impossible because of fog. Near-by were a few disconnected pools of water covered by thin ice, but short of miraculously jumping the ship from one pool to another over the intervening floes, there seemed no way for us to make progress. I glanced down our side. For several feet above the water-line, the paint was gone and our elm doubling was everywhere scraped bright with here and there a deep gouge in the wood from some jagged floe.

De Long joined me at the rail, looked despondently off through the mist, his pipe clenched between his teeth, the while he puffed vigorously away at it.

"A grand country for any man to learn patience in, chief," he remarked glumly. "Since we can't push through the pack to Wrangel Land over there on the western horizon, I've been hoping and praying at least to get the ship into Herald Island to make winter quarters before we were frozen in, but look what's happened!" He gazed over the bulwark at the near-by hummocks. "Yesterday I hoped to-day would make us an opening through to the land; to-day I hope to-morrow'll do it. And to-morrow——?" He shrugged his shoulders and left me, to climb our frosted rat-lines to the crow's-nest on the chance that from that elevation he might see over the fog. This turned out a futile effort, since not till one p.m. when the fog finally lifted, were we able to move.

With the weather clearing, I got up steam while De Long, armed with binoculars, perched himself once more in the crow's-nest, Dunbar again straddled the fore topsail yard, Chipp took the bridge, and we got under-way for as odd a bit of navigation as all my years of going to sea have ever witnessed.

To start with, the only possible opening was on the port bow, but with heavy ice ahead and astern, there was insufficient room to manœuvre the ship by backing to head her for the opening. So over the side went Bosun Cole and half the starboard watch, dragging with them one end of a six-inch hawser. Selecting a size-able ice hummock a few ship-length's off on the port side which gave a proper lead to our forecastle bitts, Cole expertly threw a clove-hitch in the hawser round the hummock, using the ice, so to speak, as a bollard; while on deck, Quartermaster Nindemann heaved in on the ship end of that line with our steam winch, warp-ing the bow smartly round to port till it pointed fair for the open-ing, when Chipp gave me the signal:

"Slow ahead!"

With a few turns of the propeller, we pushed our bow into the crack between the floes. After that, with the line cast off, it was a case of full out on the throttle. With connecting-rods, cranks, and pistons flying madly round, we certainly churned up a wild wake in that narrow lead wedging those cakes apart while the *Jeannette* squeezed herself in between the ice-floes.

And so it went for the next three hours, the captain and the ice-pilot directing from aloft, while in the engine-room we nearly tore the engines off their bed-plates and the smoking thrust-block off its foundation with all our sudden changes from "Full ahead" to "Full astern" and everything in between, while the *Jeannette* rammed, squeezed, backed, and butted her way through the ice, sometimes relying only on the engines, sometimes only on Jack Cole and his mates plodding along on the floes ahead of the ship dragging that six-inch hawser and occasionally taking a turn with it on some hummock to help warp the ship into position for ramming. Our solid bow and thick sides took a terrific beating that watch as we hammered our way through pack ice deeper than our keel, but everything held, and when we finally ceased a little after four, it was not from any fear of the consequences to the *Jeannette,* but only because the fog came down again, blotting out everything.

Once more we ran out our ice-anchor, and with that secured, recalled aboard the warping party. I came up out of the engine-room, having taken enough out of our engines in a few hours to drive us half-way to China. Chipp, Danenhower, and the captain all were gathered on the bridge over my head.

"Well, Dan, how much've we made good toward Herald Island?" I enquired eagerly of Danenhower.

The navigator's thick-set brows contracted dejectedly as he peered down at me over the after rail.

"Maybe a mile, chief," he answered.

Maybe a mile? And to get that mile, keeping up a full head of steam all the time for ramming, I had been burning coal furiously these past three hours. A hundred miles of progress at that rate and our coal would be completely gone. I turned questioningly toward the captain and asked:

"I suppose it's bank fires now and save coal, hey, brother?"

Before answering De Long looked off through the fog. Ice ahead, ice astern, ice on both beams, with only tiny disconnected patches of water showing here and there among the floes. He shook his head.

"No, chief, we won't bank this time. Let your fires die out altogether; save every pound of coal you can. If a good chance comes to move, I'll give you ample time to get steam up again."

And so we left it. As the day ended, the *Jeannette,* hemmed in by ice, lay an inert ship, unable to move in any direction, as a matter of form only, held to an ice-anchor; while below, after securing the engines, I reduced the watch to one man only, young Sharvell, coal-heaver, left to tend the boilers while the fires died out in them.

The temperature, which never during that day rose above the freezing point, started to drop toward evening and soon fell to

23°. The result was inevitable. Young ice, making during the night over all patches of open water, had by morning completely cemented together the old pack.

One look over the side in the mid-watch satisfied me there would be no call for the engines next day, nor unless something startling happened, for many a day. All the steam I could put behind my engines could not stir the *Jeannette* one inch from her bed, and as for warping her now with our winch, our stoutest hawsers would be about as useful as threads in tearing her from that grip of ice.

And so September 6, 1879, ended with the helpless *Jeannette* solidly frozen into the Arctic ice-pack.

CHAPTER X

THAT freezing into immobility of the *Jeannette* in so low a latitude, fell like an icy shower on the spirits of our wardroom mess, and from that day sociability vanished. Already Dunbar and Newcomb were not on speaking terms; Collins regarded me sullenly and the rest of the mess hardly less so; and the captain, who on leaving St. Michael's, had after an unpleasant disagreement with Mr. Collins in the ward-room, decided that he should be more punctilious and less informal in his intercourse with us, now withdrew into his official shell completely. For myself, this worried me not at all, for I well knew the effect that responsibility has on most skippers, and particularly realised (as De Long seemed finally also to have done) that for a captain not much senior in years nor in rank to most of his officers, close comradeship is incompatible with the maintenance of proper respect and authority.

However, if we had no sociability to cheer us up, we soon had plenty of other matters to make us forget the lack. The ice-pack which held us was evidently under way, headed northward, and we had not been in the pack a day before the pressure, nipping us on the beam, shoved the *Jeannette* up on a submerged tongue of ice projecting somewhere below our port bilge, giving us a list to starboard of over 5° and causing some inconvenience in getting about. As if this were not enough, after a few watches to our great uneasiness our list suddenly increased to 9°, and incidentally jammed our rudder hard starboard.

Here was cause enough for real worry. A permanent list of 9° is in itself a great nuisance in getting about on a ship even in the tropics, but now with the temperature below freezing and the decks slippery with ice, we were in a bad way to keep footing. And if the list got worse and carried away our rudder or laid us on our beam ends as it threatened to do, what then?

We promptly bestirred ourselves. Under Lieutenant Chipp's

direction, improvised torpedoes made of kegs full of black powder
were planted in the ice under our stern, but with no results. In
spite of an all-day struggle, not a torpedo could we explode. To
Chipp's intense chagrin, every fuse we had proved defective and
would not burn. And an attempt to fire the charges with that new-
fangled device, electricity, also failed, apparently because our current
was so weak it all leaked away through our non-insulated copper
wires into the ice, leaving not enough at the terminals to set off our
torpedoes.

To aggravate us while we toiled to straighten up our ship, we
had an extraordinarily clear day, giving a splendid view across the
ice to Herald Island off to the westward, with far beyond it a dis-
tinct range of peaks—Wrangel Land which, when we set out on
our expedition we had fondly expected to spend the winter exploring. Frozen in, Heaven only knows how far away from it, we
gritted our teeth and worked in the freezing weather to explode
those torpedoes, but to no purpose. Night fell and left us still in that
perilous position.

Our fourth day in the ice found us still struggling to right the
ship. The torpedoes were abandoned. We resorted to more primitive methods, those used centuries ago on sailing ships to careen
for cleaning the hulls.

Jack Cole, bosun, and a gang of seamen, swarmed up the icy
shrouds, rigged a couple of heavy tackles at the masthead, one at
the fore, the other at the main, and secured their lower ends to
ice claws hooked under the thick floes on our port side.

Then to the hoarse cry of the bosun:

"Yo, heave!" our entire crew, stretched out along the falls, lay
back and foot by foot hove them well taut, till our port shrouds
came slack and the captain signalled to belay hauling lest something carry away. But even under this terrific strain on our masts
tending to roll us to port, our vessel, gripped firmly by the ice,
righted herself not an 'inch.

De Long, regarding with keen disappointment our strained
cordage and bent masts, had still one more shot in the locker. Torpedoes had failed, careening had failed, but we had yet an ice-saw.

He motioned to Alfred Sweetman, our tall English carpenter, standing at the base of the mainmast dubiously eyeing the overstrained cross-trees above him.

"Rig that ice-saw, Sweetman!"

The carpenter responded hurriedly. While Jack Cole braced back the main yard so that its port end plumbed our quarter, Sweetman and his mates broke out from the hold our ice-saw, a huge steel blade twenty feet long and broad in proportion, its cutting edge studded with coarse teeth that would have done credit to any full-grown shark.

Under Nindemann's direction, the port watch went over the side armed with pickaxes and crowbars and started to break a hole for the saw through the ice on our quarter, while Cole and Sweetman swung the saw from a tackle at the yard-arm, weighted its lower end with a small kedge anchor, and then awaited the completion of the hole through the floe. They had several hours to wait, for not till the gang on the pack had dug down fifteen feet did a crowbar go through into the open water below, which, gushing unexpectedly upward into the hole, soaked the diggers with freezing spray and sent them madly scrambling up the rough sides of their excavation.

Fifteen feet of ice! De Long's moustaches drooped for a full due when Nindemann reported that. Only mid-September, and already fast in ice extending two feet below our keel! A gigantic block of ice to try to cut, but there was nothing for it but to saw away if we were ever to right our ship. Fortunately, our saw was at least long enough.

The bosun plumbed the hole with the kedge anchor suspended from the yard-arm, hauled everything two-blocks, and then:

"Let go!" he roared.

Down came the kedge with a run, crashed into the thin remaining ice-floor of the hole, broke through, carrying the lower end of the saw with it, and we were ready.

Then commenced four hours of strenuous labour. Sweetman and Nindemann, armed with crowbars, down on the pack, guided the sides of the saw blade for a fore and aft cut, while on deck the starboard watch stretched out along the fall, alternately heaved and slacked away, on the upstroke lifting the weight of both saw and kedge anchor, on the downstroke depending on the weight of the kedge only to drag the blade down again, while on both strokes the steel teeth rasped and shrieked and tortured our ears as they tore into the solid ice.

But it was useless. In spite of Sweetman's skilled guidance and Nindemann's brawny shoulders, it was next to impossible to keep that blade going straight against such thick ice, for the bottom of the saw being so far below them, actually guiding it was wholly out of question, with the result that on nearly every stroke the saw jammed in the cut. After half a day's arduous labour the net results were a badly bent saw, hardly a fathom of ice cut, and such a flow of sulphurous language both on deck and on the ice-pack from those handling the saw that I doubt not it may well have melted more ice than we cut.

So at eight bells, when the gang over the side knocked off for mess, De Long, ruefully contemplating the twisted saw temporarily hanging in the clear at the yard-arm, and the insignificant length of the cut compared with the stretch of ice along our hull which had yet to be severed, gave up and silently motioned Cole to unrig everything. With alacrity, all hands as soon as this was done, scrambled below to the forecastle.

A few minutes later, in the comparative warmth of the ward-room (50° instead of the 16° out on deck), with some difficulty on account of the slope, I eased myself into my chair near the head of the table on the captain's left, silently bracing my plate with my knife to keep it from sliding away to starboard while Tong Sing ladled out the soup, hardly more than half-filling my plate, lest the steaming liquid overflow the low side.

"Well, mates," observed Danenhower, at the low end of the table, contemplating his scanty portion, "such is life in the Arctic! We're in for this list all winter. I'm glad I don't like soup anyway. Stew's more in my line."

"Better see Ah Sam, then, Dan," I advised, "and make sure he thickens that stew enough to insure a safe angle of repose, or

F

your stew'll flow away like the soup."

"Don't worry, boys; I'll fix it," broke in our executive officer, Mr. Chipp. "If we can't right our ship, I can right this mess table anyhow. I'll have the carpenter saw a foot off each of the legs on the high end and that'll about compensate for the heel and level it off for us."

"Chipp, I'm ashamed of you," I objected. "Your cure's worse'n the disease. That'll fix the slope, true enough, but what'll the skipper and you and I do? Shortening these legs a foot will put this end of the table in our laps. How'll we eat then; cross-legged on the deck like a lot of Japs? Maybe you can, you must be used to it, being just back from there! But I'm afraid I'm too stiff in the joints to flemish down my legs properly!"

Chipp, who had just come back from the Orient to join the *Jeannette,* saw the point, considered a moment, then looked speculatively down the table to the low end where sat the mess treasurer and the juniors in the mess.

"You're right, chief. That'll never do; there's too much rank up here to monkey with this end of the table. Instead, I'll have Sweetman level off by adding a foot to those table legs on the starboard end."

Immediately Danenhower, facing the captain from the low end of the table, flared up.

"And what do you expect *me* to do then, Mr. Chipp? Get myself a high-chair like a damned infant so I can reach the table while I eat? And wear a bib too, maybe? Forget it!"

Chipps, squelched from both ends of the table, shrugged his shoulders.

"Well, I give up, mates! Anchor your soup plates anyway you can then. But don't be blaming me if your chow finishes in the scuppers instead of down your gullets." He relapsed into silence.

The meal proceeded with difficulty. Tong Sing, bending low over each man's shoulder in succession, sought to maintain his grip on the sloping deck the while he tried to level off the platter of salt pork long enough for each to help himself, but it was evident that it was only a matter of time till one of us got the contents of that platter in his lap. After two near accidents, avoided only by skilful juggling of the platter by the impassive Tong Sing, the captain motioned the steward to quit serving and put the dish down before him.

"Enough, Charley. Set that platter down right there. Lend a hand, Melville, at passing those plates, and I'll serve out myself. We'll have to let formality in serving go by the board till spring's here and we're on an even keel again. Let's have your plate now, Dan," he called, as the relieved steward deposited the heaping dish of pork before the captain and padded off to the galley for the potatoes.

"Spring? When does spring arrive around these parts?" asked Danenhower irrelevantly, passing his plate.

"God knows, I don't!" replied Chipp. "By June, I hope though, Why, Dan?"

"By June, eh?" The navigator counted on his fingers. "Nine months yet. And nine months more of having to navigate these careened decks is going to be tough on the legs. I'll have a permanent limp in my left leg long before that, trying to keep erect."

Dr. Ambler seated in the middle of the table looked at Danenhower, nodded seriously, and in his quiet Virginian drawl observed:

"You're right, Dan. And since we can't right the ship, we'll have to level off the crew. What Chipp just said gave me an idea. How about my amputating a couples of inches from everybody's left leg, just enough to counteract the list? That'll keep you all on an even keel."

"Hah, hah!" roared Chipp, looking at the doctor in mock amazement. "For a naval surgeon, my dear Ambler, your lack of seamanship pains me! Shorten our left legs, indeed? That's all very well for a starboard list when a man's going forward, but where'll he be when he comes about and wants to go aft? Worse than ever, with his short flipper on the wrong side! Not for me, doctor. I'll reef my legs myself on which ever side's necessary. Your idea's worse than mine!"

"I'm sunk," admitted Ambler with a grin. "So that won't work after all! And it looked such a grand scheme with a little easy surgery on the crew to avoid having to operate on all that ice!"

"If we stay here long enough," observed Newcomb, "according to that new theory which my fellow naturalist, that great English scientist Darwin, recently advanced, Nature will accommodate us to our environment. The survival of the fittest, you know."

"Well, 'Bugs,' that means we'll all ultimately become polar bears or perish," commented Ambler. "And since I don't look with pleasure to doing either, let's hope you and your biology are both wrong."

By this time, fortunately, all were served and in the ensuing attack on the salt pork, conversation languished. But in spite of the badinage about our situation and the half-humorous remedies proposed to alleviate the nuisance of for ever battling the sloping decks in working, the sloping tables in eating, and even the sloping bunks when we tried to sleep, it was evident that in the back of everyone's mind was a lurking fear of what next the ice had in store for us. And the futility of our efforts in combating the ice-pack were now too plain to all of us to sustain any further hope of effecting in the slightest degree any position our ship might assume, let alone her movements.

For some days we drifted impotently with the pack toward the north-west. With broken ice under pressure piling up along our high side and jamming our rudder hard against the pintles, the captain (who inwardly had been hoping for a series of September gales to come along and break up the pack and free us) at last reluctantly gave the order to unship it, which task with great difficulty, on account of the thick ice, Cole and Sweetman finally succeeded in accomplishing, tricing the rudder up to the davits across the stern. So the end of the first week found us a rudderless ship moving at the whim of the ice-pack, all chance of exploration gone, stopped at latitude 71° North, a latitude which had easily been reached in these same waters twenty years before by a sailing ship. And gnawing bitterly at our captain's soul was the knowledge that till summer came to free us, in spite of steam or sail, the *Jeannette* Polar Expedition must drift idly with the pack, so far from the

Pole as to be the laughing-stock of the world when it became known, the while we consumed our supplies, burned up our coal, and wore out our bodies to no purpose.

Where was the pack taking us? Anxiously we daily watched the trend of the drift-lead dropped to the bottom through a hole in the ice under our stern, then checked against occasional bearings of distant Herald Island and the few astronomical observations Danenhower got through the fogs. The navigator announced finally that we were drifting north-west with the pack, at a rate of about two miles per *day*. Where would that lead us? And when? By spring, to the shores of Wrangel Land perhaps, the captain hoped, not very optimistic apparently even for actual realisation of that prospect,

Meanwhile, I prepared for the worst below. To save coal, fires had on the day the ice caught us, been allowed to die out under the boilers. Now with our underwater hull practically sheathed in ice, the cold below was increasing, and to avoid freezing boilers and pipe-lines and bursting them as a result, it was necessary to free everything of water, leaving boilers, pumps, and engines empty, dry, and unfortunately as a consequence, unavailable for immediate service if required. Not to have some steam up and his auxiliaries, at least, ready for service, would irk any engineer. But to be even more helpless below, not even to have boilers filled and ready to light off in an emergency, gave me serious cause for worry. However, there was no way out. Keeping the water warm in the boilers and lines meant keeping fires alight which would consume precious fuel and leave us with empty bunkers when the ice at last released us and we could steam again. Keeping water in the boilers and lines without the fires, meant freezing and bursting our lines and perhaps our boilers, leaving us helpless to utilise the coal we had saved. One horn of that dilemma was as bad as the other; was ever an engineer faced with a worse choice of evils? The only way out was to be even more reckless, to empty everything, save coal, avoid freezing, and trust to luck that in a pinch somehow the boilers could be filled again with water, fired up, and steam raised once more before it was too late.

And that I did. Lee, machinist, and Bartlett, fireman, who were acting as my assistant-engineers, turned to with their wrenches. Aided by the rest of the black gang, Boyd, Lauterbach, Iversen, and Sharvell, serving as a bucket brigade, they were soon busy breaking pipe-joints, draining out water, drying out the boilers, and finally assembling everything as free of moisture as it was humanly possible to get it.

And that completed the job of reducing the *Jeannette* to a helpless hulk. No rudder with which to steer, no steam with which to move her engines, she was more helpless even than Noah's *Ark,* which indeed she soon came to resemble when the portable deckhouse we brought with us from San Francisco was finally erected.

CHAPTER XI

ON the *Jeannette,* we settled down to spending the winter in the ice-pack. The first step was to turn loose on the ice all our dogs—a proceeding greeted with yelps of joy from the dogs at no longer being prisoners, and cheers from the men who foresaw not only the prospect once again of living in some peace and safety, but also of keeping our decks clean and shipshape. There was only one drawback. Some distance from the ship we had planted bear-traps in the hope of varying our menu with fresh meat. To our disgust, instead of bear, our first catch was one of our best dogs, Smike, nipped by the foreleg between the jaws of a trap. With some difficulty, Aneguin extricated the yelping brute and the starboard wing of the bridge having been converted to a dog hospital, Smike was turned in for repairs. Hardly had this been done before a second dog, Kasmatka this time, sprang another trap and Aneguin had two patients for his canine sick-bay. This disturbed the captain, who, fearful of losing all our dogs with the sledging season coming on, ordered the traps set out only at night, when all the dogs had been herded aboard.

Meanwhile, De Long kept a watchful eye on Herald Island looming up in the distance as we drifted with the ice to the north-west. Our skipper, anxious to discover if the island contained any driftwood which might serve us for fuel, or a possible harbour if by any chance the hoped-for September gales broke up the pack and allowed us by steaming up again to reach it, determined on exploring it while still it bore abeam, apparently only five miles off. So on the captain's orders, Chipp, Alexey, and I made up an exploring party, taking a sledge, eight dogs, and provisions for a week. We set off on the morning of September 13, cheered by all the crew, and immediately I discovered something about dog-teams. Instead of my boyhood pictures of a dog-team racing in full-cry over the snow with the Eskimo driver having nothing to do except to snap a whip as he gracefully reclined on the sledge, there was chaos. The dogs yelped and fought; the leaders battling in the rear, the rear dogs in the centre, the harness all atangle, and progress the last thing apparently any dog was interested in. It took all our efforts to untangle the mess and get under-way, with Alexey whipping the dogs to hold them in line, and Chipp and I behind pushing the sledge to get it started and encourage the dogs. Fortunately for us, the going at first was fair, with much young ice, still smooth and unbroken to ease our path, but we soon ran into rough and broken floes, over which we toiled for hours. In this wise we covered fifteen miles without Herald Island appearing any closer when, to our dismay, a wide open water lead blocked our path. From the edge of that gap we scanned the island beyond, still five miles away, but clearly visible through the frosty air, to find that its shores were precipitous cliffs of rock, offering no signs of a safe harbour even if we could have worked the *Jeannette* inshore, while there was not the slightest evidence of vegetation or of any driftwood which might ease our fuel problem.

Chipp and I considered the situation. Without a boat, it was folly to attempt proceeding farther—we might, even if we managed to skirt this lead and make a landing on the island, find our return cut off by other leads and with our ship being carried to the northward by the drifting ice, be left to starve on that barren rock. Reluctantly then we turned back, but so slow was our progress over the rough ice-pack we were forced at last to camp on the ice for the night. It was not till nine the next morning, which happened to be Sunday, that we sighted the ship, a little glum at returning with nothing to show for our journey except one small seal which Alexey had shot at the edge of the lead and which we carried strapped down on the sledge.

Instead of the peaceful calm of a Sunday morning, however, I found the ship in a turmoil. As we approached the stern with our sledge, trudging wearily along in Alexey's wake and watching eagerly the thin column of smoke from the galley that to us meant just one thing—a hot breakfast—someone on deck shouted:

"Bear!"

The next I knew, down the gangway on to the ice came the quartermaster, Nindemann, a rifle in his hand, running in his stockinged feet as hard as he could toward our stem. Sure enough, there galloping off past the bow, was a big polar bear, who quickly faded from view, but that meant nothing, as white bears naturally enough do not stand out long against an ice background. A bear! Fresh meat instead of salt beef, if we got him! But polar bears had a reputation for ferocity and there was Nindemann, single-handed, going after one. What might not the bear do to him among those hummocks? Chipp and I looked at each other questioningly. Being somewhat ungainly and rather stout, I can hardly say that Nature ever designed me for chasing bears, besides which, having just tramped thirty miles across the broken pack, I hardly felt equal to joining any bear hunts, and I was about to suggest we let the Indian, Alexey, go in support, leaving us to struggle with the dogs, when the problem was solved for us. Down the gangway, going four bells in Nindemann's wake came Danenhower, also flourishing a rifle, and in no time at all after that, Collins and Newcomb, both armed, shot down the gangplank also and were off on the run. By the time our sledge made the gangway and we hauled our tired legs up the incline, not only the bear but all four hunters were out of sight among the hummocks.

As we came over the side, I looked questioningly round for the watch officer to report my return aboard, but except for Dunbar, who was already half up the foreshrouds on his way to the cross-trees, undoubtedly to get a better view of the chase, there wasn't a man in sight on deck, so without further ceremony, both Chipp and I laid below to the ward-room, where, furs and all, we planked ourselves wearily down at the mess table, calling loudly for Tong Sing and hot coffee. At the table, in no wise disturbed by the shouting on deck, was Captain De Long, still lingering over his breakfast. Eagerly he questioned us about Herald Island while we ate; his disappointment at our report, utterly dashing his hope that the island might ever serve him as a base, was plainly evident, though he tried to conceal it from us by changing the subject.

"Well, Chipp, there's still Wrangel Land to look forward to."

He gazed listlessly up at the ward-room clock. "But that'll have to wait. Right now I believe it's time for Sunday inspection. Have Nindemann muster the crew immediately on deck."

"Nindemann, sir?" asked Chipp puzzled, having just seen our quartermaster vanishing on a bear hunt.

"Yes, Nindemann of course. He has the watch now."

"Sorry, captain," answered Chipp, "but Nindemann went over the side just before we returned chasing a bear. He must be over a mile from here by now. However, now I'm back, sir, I'll muster the crew myself."

"Nindemann gone, you say? When he had the watch? Who gave him permission to leave; Danenhower, I wonder?" De Long frowned, then motioned to Tong Sing. "Tell Mr. Danenhower I want to see him right away."

"Dan's gone too, sir," put in Chipp quickly before the steward could leave. "He followed Nindemann after that bear, to back him up, I suppose. I'll arrange for the inspection, sir."

De Long's frown deepened perceptibly at this.

"So the navigator and the watch officer are both gone, eh? Who'd they leave in charge on deck?"

"Don't know, captain," answered the executive officer, "unless it might have been the ice-pilot. But Mr. Dunbar was half-way up the foremast when we came aboard, so I can't just say."

The skipper stroked his moustaches thoughtfully, finally ordered:

"Never mind the inspection, Chipp. I'll delay it till they're back. But this won't do. Even if we are in the ice, I can't have my crew disappearing from the ship whenever they see fit. Pass the word to all hands at the next muster that hereafter no officer or man leaves this vessel without first getting my permission. Do you understand that, Chipp?"

"Aye, aye, sir," replied Chipp. "Having just been across thirty miles of this infernal ice, I quite agree with you, captain. We can't have our men chasing God knows where among these hummocks and never knowing who's gone nor why. But it's not the men's fault this time, sir, it's mine. I should have covered that by an order a week ago when we entered the ice."

"Never mind that, Chipp," broke in De Long, "I'll issue the order, you just tell Danenhower and Nindemann I want to see them when they return." He rose abruptly, pulled on his fur parka, and went on deck.

Meanwhile, on the distant pack, the bear hunt was in full cry, first the bear under a full head of steam, then Nindemann tenaciously following in his wake, then Danenhower a few hundred yards astern getting somewhat winded, and finally bringing up the rear of the column, Newcomb and Collins. Over the broken ice and in and out among the hummocks ran the bear, giving his pursuers no chance for a decent shot, and all the time (by instinct, no doubt), heading away from the *Jeannette* till it was lost to sight. After fifteen minutes of hot pursuit, Danenhower, torn between the need of supporting Nindemann ahead of him and the neglected Sunday inspection behind, and disheartened also by observing that the bear was steadily gaining, stopped at last, till the rear-guard caught up with him and paused briefly at his signal.

"We haven't got a ghost of a chance now of catching that bear," panted the winded navigator to his companions, "so I'm going back to prepare the ship for inspection. But you two keep on to help Nindemann in case that bear makes a stand. Savvy?"

Collins and Newcomb, saving their breath, nodded and set off again.

Danenhower, puffing heavily, returned to the ship, hurriedly mustered the crew on deck, officers to starboard, seamen to port, and finally, an hour late, went below to report to the captain that the ship was ready for inspection.

In the chill winter morning, with the thermometer not much above zero, we stood in our furs, officers indistinguishable in those baggy garments from seamen, waiting for the captain to emerge. A bleak enough scene. Along our whole port side was broken ice, piled up by the pressure (which was heeling us to starboard) in irregular heaps till it came practically fair with the rail and threatened if the pressure increased to rise still higher and flow like a glacier down our sloping deck. Aloft, as usual, our rigging was outlined in ice, our masts and spars cased in it, and our furled sails against the yards so throughly frozen into a solid mass that had we wished to spread our canvas, it would have been beyond the power of human hands even with axes to loose one fold from another. That made me smile a bit. Our sails were even more useless than our engines, for given time, I could at least fire up again; but I could see no way in which Chipp could possibly make sail till summer came once more.

In the midst of my meditations, De Long emerged from the poop. Swiftly Danenhower called the roll, saluted the captain, reported three men absent. The captain, to whom by now this was no news, acknowledged curtly. The crew was dismissed, fell out, went below, and stood by their various stations there while the skipper inspected the berth deck, the galley, the store-rooms, and, in short, every space and hold, commenting briefly now and then. As a whole, the ship was dry; in spite of the cold outside, no condensation and no frost as yet showed in our living quarters or store-rooms.

Inspection over, the bosun passed the word for Divine Service in the cabin aft, but except for the officers aboard, the captain's congregation was small. Attendance being voluntary, the majority of the crew stayed away, which may perhaps have been taken as a good omen, for I well believe the old saying to be so, that the reliance of a sailor in God is in inverse proportion to his faith in the strength of his ship. Evidently, then, our seamen, seeing no special danger in our predicament, felt no great need for prayer, leaving that to the captain, who, they well knew, was in the absence of a chaplain required by the Navy Regulations to hold services. I may say here, however, for Captain De Long that he was a deeply religious man and it required no compulsion from the Regulations in his case to insure Divine Service and to ask the blessing of the Almighty upon his undertaking and his crew. Personally, however, his position in the matter was a little odd, because De Long himself was of the Catholic faith. Nevertheless, since most of the crew were, if anything, Protestants, he always conducted the services in the Episcopal ritual.

Services over, we laid up on deck, to find practically the whole crew lining the rail watching the absent hunters straggling back across the ice-pack to the ship, with the winded Nindemann leading and the other two well in his rear. There were no signs of the bear, who had evidently successfully outrun his pursuers, but what I think mostly engaged the seamen's attention as their eyes moved covertly from the captain at the starboard gangway to the returning hunters on the ice was their expectation of seeing the skipper light into the absentees for missing inspection. Nindemann, first over the side, a little surprised at seeing the captain at the gangway instead of Dunbar to whom he had turned over the deck, saluted De Long, reported casually:

"Returning aboard, sir," and unslinging his rifle, turned to go below.

"Wait a minute, Nindemann." The quartermaster paused, De Long eyed him silently a moment, while the crew, a little forward, looked eagerly for fireworks, but to their disappointment De Long said very quietly:

"Nindemann, you were watch officer. For a quartermaster with all your years at sea, I thought you knew better than to leave the ship without permission. Never let it happen again."

Nindemann, his stolid German countenance flushing under even that mild reproof, hesitated a moment between the relative desirabilities of silence and justification, then muttered:

"But, cap'n, before yet I go, I turn over the deck to Mr. Dunbar. And that bear, he was already yet running away. There was not time for anything."

De Long shook his head.

"In an emergency, Nindemann, a watch officer may turn over the deck and leave. But a polar bear is *not* an emergency. Don't do it again. That's all now. Go below."

For an instant, hoping to explain further, the quartermaster hesitated, but one glance into De Long's quiet blue eyes changed his mind.

"Aye, aye, sir." He gripped his rifle, shuffled forward past his shipmates.

By this time, Collins and Newcomb were coming up the gangway. The knot of sailors, disappointed in the expected scene over Nindemann, lost interest and scattered. If the captain would not blow up a seaman for a serious breach of discipline, he would hardly lay out an officer for less. And in this they were correct. De Long went below before the two hunters reached the side; they reported their return to Dunbar, and had not Danenhower stopped them, might have laid below unhindered. But the navigator, curious as to events, laid a brawny arm on little Newcomb's shoulder, asked the naturalist banteringly:

"Well, 'Bugs,' how did you make out with that specimen of *Ursus Polaris?*"

"*Ursus Polaris?* There is no such specimen. *Thalassarctus maritimus,* you mean," blandly replied Newcomb. "I regret to say the specimen outfooted us, and neither the quartermaster, the meteorologist, nor I unfortunately got in a shot."

"Too bad," agreed Danenhower, "but what, by the way, is a thalassa—— What did you say?"

"*Thalassarctus maritimus,*" repeated Newcomb. "What the untutored call a polar bear or in Latin, *Ursus Polaris.* That's all wrong. It's an ice bear or, technically, a *thalassarctus maritimus.*"

"Well, well!" grinned Danenhower, "marvellous how a bear weighted down with a name like that can run, isn't it? By the way, 'Bugs,' when you've stowed your rifle, you'll have a chance to show off your Latin to the skipper. He wants to see you in his cabin." He turned from Newcomb to the panting meteorologist. "And a little later he'd like to see you too, Collins."

"Me? About what?" demanded Collins sharply.

"Just a little private warning about leaving the ship without permission, I guess. He's already reprimanded me for it." Danenhower laughed. "My fault, of course. I should have known better."

"Well, I *shouldn't,*" snapped out Collins, and disappeared through the door in the poop bulkhead, leaving Danenhower looking after him, amazed at the heat of his reply.

Sunday dinner was a quiet meal in spite of the fact that in the cabin for our main dish we had an unusual treat—roast seal—the one that Alexey had shot on our trip to Herald Island, and which we had dragged back on our sledge. The seal meat was excellent, something like rabbit, I thought, and a very welcome change from salted beef and pork, but nevertheless, except for Danenhower chaffing Newcomb, there was little conversation. I was tired from the journey to Herald Island, so also I knew was Chipp, but the wet blanket on the conversation was evidently Collins, who, mum as a clam, sat through the meal without a word to anyone, and as soon as he had cleaned his plate, departed suddenly without a "By your leave" to anyone.

De Long, a little perplexed at Collins's quick departure, hastily drew a paper from his pocket, and broke the silence.

"Gentlemen, before anyone else leaves, here is an order I've issued to prevent a repetition of what happened this morning. Each of you please read and initial it."

The order passed rapidly round the table. It was brief enough, requiring each officer and man to get the captain's permission before leaving the ship, and requiring him to report both his going and returning to the officer of the deck. When all had noted and initialled it, the captain called to Tong Sing:

"Charley, show this immediately to Mr. Collins, and tell him to initial it."

Tong Sing took the order, padded placidly out of the cabin in search of our departed messmate.

A little later, I went on deck myself. There outboard of the fore-mast, leaning on the port rail, morosely watching the pack of dogs on the ice snarling and fighting over the scraps of seal which Ah Sam had flung them, was Collins. From his flushed face and his agitated manner it was evident our meteorologist was much upset. While Collins's puns had always much annoyed me, and my casual jokes had no doubt irritated him, still we were friends, and on my appearance from the poop, he beckoned me to join him, which I did.

"I'm trapped, chief!" he burst out heavily. "Back in the States, my brother warned me I shouldn't have shipped on this cruise as a seaman, but like a fool, I didn't believe him then! Now it's happened, and I'm trapped!"

"You trapped? What's ailing you, Collins?" I asked, astonished at this hysterical outburst. "We're all trapped with the *Jeannette* in the ice, but you're no worse off than I am."

"It's not the ice, chief!" Collins gripped my arm, drew me close to the rail. "It's the captain! I've been fearing this for weeks. You're all right, you're an officer. But I was fooled into shipping as a seaman! Now the captain's got me where he wants me. Look at that!" He reached inside his parka. I looked. "That" was a somewhat faded newspaper clipping of an interview De Long had given a reporter from the *Washington Post,* an interview which months before I had once seen reprinted in a San Francisco newspaper.

"That's where it started; look what De Long called me there!" With a shaking finger Collins pointed to the middle of the clipping, in a voice quivering with emotion, read an extract:

"'It may be that some specialists or scientists will be invited or permitted to accompany us, but they will be simply accessories.'"

"See that? *Accessories!*" Collins's voice choked. "He's labelled me as simply an *accessory,* Melville! I should have quit as my brother advised me when I first saw that interview, not gone and shoved my head into a trap by signing as a seaman!"

I looked at him curiously. Undoubtedly the man was overwrought, seething with suppressed passion which something had finally touched off. I tried to calm him.

"Now see here, Collins, what are you taking offence at? What's so bad about your being an accessory to the Navy in this scientific stuff? You don't think, do you, that in this expedition De Long and the whole Navy should be accessories to you?"

But Collins, boiling inwardly, did not even hear me. He seized my arm again, continued vehemently:

"And now he's sprung his trap. That order he just sent me to sign! And to show what he thinks of me, he picks his Chinaman to order me to sign it! I'm in his power, on the books as just a seaman! Fool! If I'd had a grain of sense, I shouldn't have come except as an officer or at least as a passenger!"

Collins was certainly beside himself. I looked swiftly round, fearing he would make himself ridiculous before the crew, but fortunately they were all still below, lingering over their Sunday dinner. I turned back to Collins.

"But what's bothering you, anyway, brother? What's this trap you're so excited about?"

"Don't you see it, chief? It's plain enough. I'm only a common seaman here. In the captain's power! And now to humiliate me, he's forbidden me to leave the ship without begging his permission!"

I stared at Collins incredulously. Was that all? If it had not been for his overwrought features, I could have laughed in his face.

"Don't be so damned morbid, Collins," I replied as gently as I could. "About that seaman business, you're as much an officer aboard this ship as I am, regardless of how the law required them to put you down on the ship's articles. Don't you live in the cabin, mess with the officers, muster with the officers? What more do you want? Some gold lace on your sleeves? But even if you rated it, what good would it do you? Not one of us wears it here. As for the captain's order, it hits me and every officer and man aboard as much as it does you. It's just part of the ship's discipline."

"Ship's discipline! Oh no! That order's aimed at me, personally! To make me beg for every little right. To take away my liberty. Because he fooled me into signing on as a seaman, the captain thinks now he can take away my rights. But I'll show him! He can't persecute me!"

Here was a damned mess. Hardly ten days in the ice and our meteorologist already talking insanely about persecution. He had the civilian's foolish idea that aboard ship by some hocus-pocus an officer was a god, a passenger a free agent, and a seaman but a slave. Didn't he realise by now that in the Navy every man aboard ship was equally subject to the captain's authority; that in the hands of a tyrannical captain, an officer's stripes afforded no protection from abuse? That if the captain really wished to humiliate and persecute him, a commission as an officer could not possibly save him? I tried to calm Collins's fears.

"That order's innocent enough, Collins, and it's meant for all hands. The skipper never thought of you when he wrote it."

G

"Oh, yes he did! It's aimed at me, all right. But I'll fool him!" Collins's eyes positively glittered with rage. "Try to make me beg his permission, huh? I'll start a silent protest by staying aboard. Before I ask De Long's permission to leave, I'll not go off this ship again even if I die for it!"

I gazed at Collins in perplexity. An impulsive Irishman if there ever was one, going off half cocked over a perfectly innocent order. What ailed the man? Did he think the captain was jealous of his professional attainments? was he afraid the captain meant to prevent him or anyone else aboard from reaping what glory he might from the success of our expedition? That outburst about being called an accessory—what suppressed emotions did that reveal? Was Collins such an idiot as to think that De Long after years of fighting and sweating to make this expedition a reality, was now going to act merely as sailing master on his own ship, putting aside his own dreams and ambitions of discovery in favour of a minor assistant of whose very existence he had been ignorant till a few short months before? I would never have believed such egotism possible, but as I looked into Collins's distorted face, I began to wonder. However, so far as I was concerned, that was neither here nor there. We were going to have a long time in the ice yet together, and if life was to continue reasonably pleasant in the imprisoned *Jeannette's* cabin, Collins must not make a fool of himself.

"Come now, Collins," I begged persuasively, "think it over, and you'll see what I tell you is so—the order's reasonable enough. But even if it weren't, you'd only make a bad matter worse by your 'silent protest.' I wouldn't do that. It bears on me the same as it does on you. Now I'm an officer of twenty-three years seniority, which is more than De Long has, and were we both on board a frigate, I'd be very much Mr. De Long's senior. But here on the *Jeannette* he's captain and my superior, so I don't feel it bears on me at all that I have to ask his permission to come or go—it's only a custom of the Service. And there's the skipper now," I added as De Long appeared on deck from the poop and stood blinking a moment in the glare of the ice. "Think it over!"

But unfortunately for my clumsy efforts to pour oil in the troubled waters, Collins's eyes, gazing out over the ice, happened to fall at that moment on the two little wood and canvas outhouses a ship's length off the starboard beam, which served officers and men as toilets, since frozen in as we were, the regular ship's "heads" on the *Jeannette* itself had been placed out of commission. To these "heads" on the ice all hands, of course, went freely as nature called. Collins's eyes lighted up as he contemplated them. He faced me with a queer grin.

"Well, chief, I'll modify a bit what I just said about asking permission to leave the ship. In such simple language that he can't possibly misunderstand, I'll beg the captain's royal permission every time I have to visit the 'head' and I'm going to start right now!" He turned aft toward the poop.

Amazed at Collins's intended action, I grabbed his arm and stopped him short.

"Look here, old man, none of that! Do you want to insult the captain openly?"

Collins twisted out of my grip.

"What do you think he's trying to do to me, chief? I'll merely be carefully obeying his order. By God, I'm going to ask him to let me go on the ice right now!" He strode aft, stopped before the skipper, saluted him elaborately.

What he said to De Long, I can only imagine, since I was too far away to hear, but I judge he phrased his request in about as plain old-fashioned Anglo-Saxon as it could be put, for De Long, obviously startled, flushed a fiery red, retorted angrily, and then turned on his heel.

And from that time forth, Mr. Collins and Captain De Long remained separate in all things as much as they could, simply carrying on the duties of the ship. And from that time also, Collins, fancying offence to himself in almost every remark made in the ward-room mess, withdrew more and more from association with the rest of us, sticking only the more closely to Newcomb, who as the sole other non-sea-going civilian aboard, he may have considered as a sort of fellow-victim.

CHAPTER XII

SEPTEMBER passed, and the hoped-for gales which might break up the pack and allow us to escape, or at least to work into a winter harbour in distant Wrangel Island, failed to materialise. October came and went in the same manner, no real gales, no winds strong enough to have any effect on the ice, nothing but daily gusts of fine snow which cut our faces and spoiled our footing for exercise. Frozen in, we went with the ice-drift, in a general north-westerly direction, till the rocky outline of Herald Island faded into the hummocky horizon to the south, while our continued failure to sight land to the westward made it less and less likely every day that Wrangel Land stretched north-west, as we had been led to expect. But we were not idle. After all, our expedition was a scientific one. Aside from attempting to reach the Pole, aside from discovering new lands in this unexplored ocean, our major aim was to add to the world's knowledge of the Arctic seas, of the Arctic skies, of magnetic phenomena, of meteorological information and of animal life in the unknown north. For these purposes we were the most elaborately equipped expedition which had ever gone north. We carried two scientists, and God only knew what varieties of scientific instruments gathered from the Smithsonian Institution and the Naval Hydrographic Office.

Since exploration and discovery were for the present out of question, De Long turned to all hands intensively on these scientific phases. On the ice a hundred yards from the ship so as to be unaffected by the iron in her we set up a canvas observatory, with compass, dip circle, anemometer, rain-gauge, barometer, pendulum, and a variety of thermometers. Over the side, through a hole chopped in the thick ice, we provided an opening for our dredge and our drift-lead. Hourly we took observations (and carefully recorded them) of every type of phenomenon for which we were equipped to measure—magnetic variation and dip, wind velocity and direction, humidity, air pressure and temperature, gravity readings, temperature of the sea at top, bottom, and points in between, salinity of the sea water, speed and direction of drift—all this data laboriously read night and day in the Arctic chill went into our logs. And for the zoological and botanical side of our expedition, all hands were directed to bring in for Newcomb's inspection specimens of anything found on the ice, under, or above it, which meant that whatever our guns could knock down in the form of birds or beasts, or our hooks could catch in the way of fish, passed under Newcomb's scrutiny before (in most cases) they went to Ah Sam and were popped into the galley kettles.

And to top off all in completing our polar records, we brought along an extensive and expensive photograph outfit, intending to get a continuous record of our life in the Arctic and particularly some authentic views of Aurora Borealis.

So there being nothing else to compete with it for our time, science received a double dose of attention, too much, in fact. Taking the multitude of readings every hour (there were sixteen thermometers alone to be read) kept the watch officer hopping, and as each of us, except Collins and Newcomb, had ship and personnel matters to look after, it became to a high degree a nuisance. Most of this scientific work naturally should have fallen to Collins and Newcomb, but unfortunately matters in their departments went none too smoothly. The captain received a severe jolt when he leraned that the photographic outfit, entrusted to Collins's care was, practically useless because our meteorologist had neglected when buying his photographic plates in San Francisco to get any developer for them, and that not a picture he took could be developed till we got back to civilisation. When on top of this, one of our barometers and some of our precious thermometers entrusted also to Collins were carelessly broken, the captain began to mistrust Collins as a scientist and loaded a considerable part of the observation work on Chipp, on Ambler, and on me—a development which did not help to make any more amicable the attitude of Collins towards his shipmates.

Speaking frankly, after two months' close association in the cabin of the *Jeannette,* we were beginning to get tired of each other's company. Life on shipboard is difficult at best with the same faces at every meal, the same idiosyncrasies constantly rubbing your nerves, the same shortcomings of your messmates to irritate you; but ordinarily there are compensations. Shore leave gets you away from your shipmates, while foreign ports, foreign customs, foreign scenes, and foreigners give flavour to a cruise that makes life not only livable, but to my mind rich in variety, and to a person like myself, completely satisfying. But in the polar ice, we came quickly to the realisation that life on the *Jeannette* was life on shipboard at its worst—a small cramped ship, a captain who socially had retired into himself, only a few officers, and not a solitary compensation. No possibility of shore leave, no foreign ports—nothing but the limitless ice-pack holding us helpless and no hope of any change (except for the worse) till summer came and released us. And, impossible to conceal, a mental despondency, as ponderable and as easily sensed as the cold pervading the ship gripped our captain as we drifted impotently with the pack between Herald Island and Wrangel Land, a thousand miles from that Pole which in a blare of publicity from the *Herald,* he had set out in such confidence to conquer.

Gone now were all the fine theories about the Kuro-Si-Wo Current and the open path to the northward through the Arctic Ocean that its warm waters would provide. We had only to look over the side at the ice-floes fifteen feet thick gripping our hull to know that the "black tide" of Japan had no more contact with these frozen seas than had the green waters of the Nile. And just as thoroughly exploded was that other delusion on which we had based our choice of route—the Herr Doktor Petermann's thesis that Wrangel Land was a continent stretching northward toward the Pole along the coasts of which with our dog teams we could sledge our way over firm ground to the Pole. Every glimpse we got of it as we drifted north-west with the pack for our first eight weeks showed conclusively enough that Wrangel Land was nothing more than a mountainous island to the southward and not a very large island at that. As for Dr. Petermann and his idea that Greenland stretched upward across the Pole to reappear on the Siberian side as Wrangel Land, if that ponderous German scientist who so dominated current European opinion on polar matters could have been forced to spend a week in our crow's-nest observing how insignificant a speck his much-publicised Wrangel Land formed of the Arctic scene, I am sure the result would have been such a deflation of his ego and his reputation as might be of great benefit at least to future explorers even if too late to be of service to us in the *Jeannette,* already led astray by the good doctor's teachings.

How much the general knowledge amongst our officers that every theory on which the expedition had been based was false had to do with the lack of sociability and of harmony among us, and how much of it may have been owing simply to our physical imprisonment in the ice, I will not venture to say. But in my mind, the belief of all that as a polar-exploring expedition, we were already a failure, doomed never to get anywhere near the Pole, had a decided, if an unconscious, bearing on the reactions of all of us, and most of all on the captain and on Collins, both of whom had brought along massive blank journals whose pages they had confidently expected to fill with the records of their discoveries.

The captain's journal I sometimes saw, as each evening around midnight he toiled over his entries. Instead of records of new lands discovered, of the attainment of ever-increasing latitudes exceeding those around 83° North reached by the English through Baffin Bay and Smith Sound, how it must have gnawed the captain's heart that his entries had to be confided to such items as my struggles with our distilling apparatus, our difficulties with such newfangled gadgets as telephones and electric generators, or the momentous facts that Aneguin, Alexey, or Captain Dunbar (as the case might be) had chased a polar bear (or perhaps a walrus) which had been shot (or had escaped). All of these happenings to De Long's chagrin must be recorded as having occurred in the low seventies, latitudes far to the south of those reached even by the insignificant and ill-equipped caravels of Dutch seamen three hundred years ago in their explorations of Spitzbergen.

What Collins put in his journal, I never knew. But I can well imagine how much it must have irked him, a newspaper man accustomed to live in an atmosphere of printing presses rumbling away over their grist of momentous world events to be spread daily before the eager eyes of *Herald* readers, to have nothing to record except perhaps his personal sense of injustice. Yet put down something every day he did, for I can still see him, his long drooping moustaches almost sweeping the pages, religiously bending over the leather-bound ledger every afternoon in his chilly cabin in the *Jeannette's* poop, pouring the bitterness of his soul on to those pages, building up a record with which I doubt not he hoped when we returned to civilisation to blast De Long out of the Service in disgrace.

CHAPTER XIII

ON November 6th, two months to a day of our being trapped in the pack, came the first break in the monotony of our imprisonment. About four in the afternoon Collins, trudging perhaps for the thousandth time the rough path to the observatory across that hundred yards of ice which we had come to regard as substantial as a Broadway sidewalk, came pell-mell back to the ship and up the gangway into the ward-room to startle us with the news that the pack ice had cracked wide open between our ship and the observatory! We rushed on deck and over the side. Sure enough it was so. A little behind Dr. Ambler and the captain, I arrived at the edge of the rent, over a yard wide already and continuously growing wider. While we could still jump the gap, there was a wild dash to get our precious instruments out of the observatory and back across the opening to the ship, which (all the officers taking a hand) we shortly accomplished without mishap. That done, with varying emotions we watched as over the next few hours the chasm widened, with the dark sea water showing in strong contrast to the whiteness of the snow-covered ice. But not for long did we see really open water, for with the temperature far below zero, the water which was welling up to within two feet of the top of the parted edges of the floe promptly froze, even though it was salt, into a sheet of young ice. The gap nevertheless kept widening till by midnight it was perhaps ten fathoms across.

What was causing the rupture? One man's guess was as good as another's, and all were worthless, I suppose. There was little wind, no land in sight for the edge of the pack to strand on, no evidence of pressure from any direction, and plenty of water beneath us, for the soundings showed over twenty fathoms to a soft mud bottom. Chipp's surmise, that a tidal action was responsible, was as good an explanation as any. But what is not satisfactorily explainable is always fearsome, and it was perhaps excusable that we looked with some anxiety toward our ship and were secretly relieved to see her as steady as Gibraltar there in the ice some fifty fathoms off, still heeled as usual to starboard with her masts and spars showing not even a quiver as they stood sharply outlined against the frosty polar sky. And so the day ended.

But morning brought a different scene. During the night from somewhere came a push on the pack which closed that chasm, forcing the layer of young ice which had formed over it up into broken masses on our floe. Then with all the young ice squeezed out, the two parted edges of the original pack came together under such great pressure that the advancing sheet was shoved up over the edge of the floe holding the ship, leaving broken masses seven to eight feet thick strewn helter-skelter in a long ridge along the line of junction.

As an engineer, I regarded that broken ice with severe misgivings. We fortunately were solidly frozen in, with our thick floe spreading in all directions interposed as a buckler between us and the pressing pack, but suppose our floe should split and leave us exposed? Could any ship withstand a squeeze in that Titan's nutcracker? In spite of our thick sides and reinforcing trusses, the sight of those eight-foot-thick blocks of ice tumbled upon our floe was not reassuring.

On the *Jeannette,* men and officers alike questioningly scanned the scene while slowly the hours drifted by and we waited apprehensively in the silence of that Arctic morning for what was next, and while we waited even what light breeze there was died away to a perfect calm. Then without apparent reason and without warning, the gap in the ice suddenly yawned open to a width of some five fathoms and immediately down the canal thus formed, broken ice started to flow in a groaning, shrieking mass that so shook the floe in which the *Jeannette* was embedded that to us there, only a few yards away clinging to the rail of our ship, it appeared each instant the sheet of ice protecting us must shatter and the *Jeannette* herself be sucked in to join that swirling maelstrom of hurtling ice-cakes. Our eyes glued to the quaking floe into which we were frozen, we watched it shiver and throb under the battering of the broken blocks hurrying by, inwardly speculating on how long it would stand up. Occasionally I glanced furtively at the five sledges standing on the poop, packed with over a month's provisions for men and dogs, ready at a moment's notice to go over the side should we have to abandon ship. But if our ship, torn loose and caught in that mass of churning ice, was crushed and sank, how could we ever get safely away from her with our lives, let alone get clear those sledges carrying the food?

Five hours of that scene and of such thoughts we stood, and then, thank God, the flow of ice stopped. The *Jeannette* was unharmed. We were still safe. But how long a respite would we have? Who knew? Evidently not our captain. As I went below, worn and frozen, I heard him call out to our executive officer:

"Knock off all regular ship's work, Chipp. Turn to immediately with all hands and make a couple of husky sledges to carry our dinghies over that ice if we have to abandon ship. And for God's sake, shake it up!"

We got a day's rest, if one may call it that, while Nindemann, Sweetman, and both watches toiled feverishly on the sledges. Then came another day of strain, watching the moving ice grinding and smashing at our floe, breaking it away to within a hundred feet of us. Then a brief respite overnight, only at 6 a.m. to have the motion start again worse than ever.

This time, hell seemed to have broken loose. From the pack came a noise the like of which I never heard before on land or sea, in war or peace, sounding like the shrieking of a thousand steamer whistles, the thunder of heavy artillery, the roaring of a hurricane, and the crash of collapsing houses all blended together as down that canal in the pack, a terrifying sight to behold, came stupendous pieces of floe ice as high as two-and three-storey buildings. Sliding by crazily up-ended, they churned and battered against each other and against the thick edges of our floe with such unearthly screeching and horrible groanings that my ear-drums seemed in a fair way to split under the impact of that sound!

Occasionally a berg would jam in the canal, blocking the current. With that, under the force of the ice pressing behind, our floe would groan and heave up into waves till several feet of its edge cracked off, easing the pressure and relieving the jam—but each time leaving us with less and less of the floe between us and disaster.

Half an hour of this in the dim light of the early dawn, and then the movement ceased, leaving our tortured ears and jumping nerves to return to normal as best they could while the day broke. But our relief was considerably temperered when in the better light we discovered that a new crack had formed a little distance ahead across our bows and that into this opening an advancing floeberg was being driven along like a wedge towards our port side, threatening to cut into the undisturbed pack there and leave us embedded in a tiny island of ice, to be exposed then to the wear of churning bergs on both sides of us!

With no further noticeable movement of the pack, we were left in peace to contemplate the possibilities of this situation till late afternoon, when the main stream again got under way and bombarded our floe to starboard heavily for four hours so strenuously that it seemed to all of us that this time we must surely go adrift. But at about 8 p.m., the motion ceased again, leaving us all in such a state of mind that the captain's order for all hands to sleep in their clothes with knapsacks close at hand ready for instant flight, seemed to us the most natural thing in the world.

We didn't get much sleep. Hardly had the mid-watch ended, when little Newcomb, who unable to rest at all, had in spite of the bitter cold stayed on deck till 4 a.m., darted into De Long's cabin, seized his shoulder, woke him with a shout:

"Turn out, captain! It's all over this time! That ice is coming right down on us!"

De Long, already fully clothed, sprang from his bunk, seized his knapsack, and rushed on deck. The rest of us in the poop, none too sound asleep ourselves, were roused by the noise and hurriedly followed him up to find that Newcomb had hardly exaggerated.

On the starboard side, like buildings being poured through a chute, the broken floes were cascading along the channel at a livelier rate than ever, but that at least was hardly novel to us now. What froze our blood as we stood there in the cold light of the moon was the sight ahead. The rift in the pack which yesterday was headed across our bows, had changed direction squarely for our bowsprit, and now along that opening was coming toward us irresistibly and steadily, towering as high as our yard-arms, a torrent of floebergs, thundering down on the yet unbroken pack between with a violence that made the sturdy *Jeannette* quiver under our feet like jelly!

Hardly audible in the roaring of the ice, Jack Cole shrilled away on his bosun's pipe, then his hoarse voice bellowed along the berth deck:

"All hands! Stand by to abandon ship!"

Our entire crew poured up from below to shiver in a temperature of twenty below zero and shake, I have no doubt for other good reasons, as they stood helpless round the mainmast, all eyes riveted on that fearful wall of advancing ice, with a crest of hummocks, weighing twenty to fifty tons each, toppling forward like surf breaking on our floe. Another crash, another startling advance of the floebergs, and on top of the deck-house I saw De Long suddenly grasp the mainstay with both hands and hang on for dear life, awaiting the final smash as that Niagara of ice struck us.

The blow never came. God alone knows why, but hardly twenty-five feet from our bows, the onrushing wall of ice suddenly halted, the pressure vanished, and we on the *Jeannette* were left to contemplate, in the deathly Arctic silence which ensued and in the growing light, the indescribable wreckage that had been wrought in the level floe that had once surrounded us. And then like a feeble anti-climax, the stillness was broken by the whistling of the bosun's pipe, followed by his call:

"All hands! Lay below for breakfast!"

Breakfast? Who really wanted breakfast? What each of us earnestly wished was only to be far to the south, away from that dreaded pack ready to crush us, but seemingly delaying the fatal moment as a cat delays, knowing that the mouse with which it toys cannot get away.

CHAPTER XIV

NOTHING else happened that day. Our dogs, which in the face of disaster we had rounded up and penned inside the bulwarks, where they relieved themselves by staging a continuous battle, we now let loose, and they joyously celebrated their freedom in chasing each other over the broken ice. Watching their antics was some relief, little though it might be, to frayed nerves, helped to take our imaginations off what that broken ice threatened to our ship.

As a further distraction, we had a clear day and far to the southward sighted mountains, which we made out to be the familiar north coast of Wrangel Land, some sixty miles away.

And that was all. The day which for us had dawned in imminent peril, ended quietly with the *Jeannette* still frozen in that two-months'-old cradle of ice, still uncomfortably heeled well over to starboard. We began to breathe more freely.

We took no more meteorological observations, but so far as I was concerned, I had more to do than before. Though the fires were out in my boilers and all the machinery laid up, at De Long's direction I spent a great part of my time below during this period continually scanning the sides and the trusses for any signs of giving way, and inspecting the bilges to see if the ship was making any water. Of such troubles there were no indications, but I had constantly while below to be wary of my head, for I found that the banging of the ice shook down a good deal of loose matter in the holds, and particularly in the bunkers.

November 13th, one week from the day the pack first opened up on us and inaugurated our reign of terror, brought new excitement. Sleeping as before in my clothes, I was wakened at 2 a.m. by a loud crack which seemed to come directly from our keel. I slid from my bunk, in the passage outside bumped into the captain, and together we ran on deck, there to meet Collins who, on the mid-watch taking the hourly temperature readings, had rushed over the side and now was coming back aboard. He reported that there was nothing new except a crack in the ice not over an inch wide running out from our stem. This was disquieting, but nothing else happened during the night and the daylight hours passed quietly enough without further disturbance; so much so that by afternoon the skipper (full of scientific zeal and expecting apparently some days of peace) ordered our meteorological instruments be reinstalled in a temporary observatory. This we accordingly erected on one of the newly-formed hills of ice as far from the ship as we dared, but still fairly close aboard our starboard side.

From the grumbling of the seamen at this task as they dug into the flint-like ice for anchorage for the guys holding down the canvas tent over the instruments, I would say that the captain's optimism was hardly shared by his crew, but that was neither here nor there, and by 5 p.m. when the twilight faded and night fell, the job was done.

Chipp took the first sets of readings. At eight o'clock, after supper, I relieved him, to trek over the broken ice, and by the dim light of an oil lantern inside that flapping tent, read the dip circle, the barometer, the anemometer, and a varied assortment of thermometers. All the time as I struggled for footing on the rough ice pinnacle I wondered what earthly good it all was, considering the negligible chances of any of this data ever being returned home for scientific minds to study.

At 10 p.m., I turned the job over to the captain (who, staying up, anyway, the while he wrote in his journal, ordinarily took the readings till midnight when Collins relieved him), and as was now my habit after a week of alarms, I turned in with my fur boots on, earnestly hoping to get some sleep to make up for the past week's wear and tear.

Till 11 p.m., De Long in his cabin scratched away industriously at his journal. Then six bells struck, he dropped his pen, drew on his parka, and went over the side to take the hourly observations.

Being the commanding officer, and not one of his subordinates (in whom such an appreciation of the beauties of nature at the expense of punctuality in observations might have seemed a fault), De Long, on his way to the observatory, paused a few moments to stand on an ice hummock and admire a splendid auroral exhibition, a magnificent prismatic arch to the northward, filling the sky from east to west and reaching almost to the zenith. The beauty of this phenomenon was no longer a novelty to any of us, but still he stood awe-struck in the silent night drinking in that soundless electrical play of coloured light, when he heard behind him a crisp crackling as one of our dogs walking on the snow. Turning, he saw to his surprise no dog but instead two men, our so-called "anchor watch," racing down the starboard gangway and over the ice to our stem.

Both the aurora and the still unread instruments were forgotten as the captain ran immediately for the bow. To his astonishment, there he found the ice-pack peaceably floating away from our port side, leaving it completely exposed with open water lapping our hull for the first time in months!

And as he watched in dazed amazement, the gap opened, so that in a few minutes we had alongside the *Jeannette* thirty fathoms of rippling water in which was gorgeously reflected the northern lights (a detail the beauty of which I think our captain now took little note). The split in the pack was as clean and as straight along our fore and aft centre-line as if a giant hand had cut the ice with our keel, leaving the ship still embedded in the starboard floe toward which she heeled. Meanwhile the port side pack, intact even to the bank of snow which had built up above our gunwales, was sliding noiselessly away to the northward, carrying with it, still asleep, three of our dogs who had bedded themselves in its white crust!

A glimpse at our heeled over clipper bow and at our bowsprit thrusting forward over his head, quickened De Long into action. Nothing visible now remained to hold that tilted ship from sliding any second out of her bed and into open water! Back aboard he rushed, and once more the quiet of the night was torn by the whistling of the bosun's pipe and Cole's hoarse cry:

"All hands! Shake a leg! On deck wid yez!"

And again no sleep, as hastily in the darkness we hurried our meteorological instruments back aboard, struck the observatory we had so laboriously rigged only a few hours before, chased on board all the dogs we could catch, rigged out our dinghies and our other boats for immediate lowering, dug our steam-cutter out of the ice alongside and hoisted it aboard, ran in our gangway, and lastly rigged out a fall for lowering provisions over the side and into the boats.

That all this, on the sloping deck of the *Jeannette,* was done in the darkness at fifteen below zero and completed by midnight in less than an hour, indicates what speed and strength fear gave to our fingers and our feet. For the men tumbling up from below had to look but once at the precarious perch to which the *Jeannette* clung to send them flying to their tasks.

Midnight came.

Our work done, we stood by in the inhuman cold momentarily expecting to feel the ship lurch under our feet, slide suddenly off into the water, and without rudder, without steam, and without sails go adrift in the darkness in that ever-widening rift in the parted pack.

After an hour of this, with nothing happening to relieve the strain, the tension became almost unbearable. De Long, looking over the silent groups of fur-clad seamen clustered there on deck alongside the boats, ordered Ah Sam to fire up the galley range and serve out hot coffee to the men, hot tea to the officers. He then told Cole to pipe down, but with all hands to stay in their clothes, ready for any call. So we lay below, but I doubt if anyone had much better luck than I getting to sleep again.

There was no need for reveille in the morning. The first streaks of light found the whole crew from Irish bosun to Chinese cook lining the bulwark, staring off to port. I climbed the bridge to get clear of the snarling dogs. There before me, already ensconced in the port wing was the skipper, rubbing his glasses to clear them of frost for a better view.

"What do you make of it, chief?" asked De Long, nodding in the direction of the distant pack.

I squinted off to port. A thin skin of young ice, possibly four inches thick, had formed over the exposed water. Across that, per-

H

haps five hundred to a thousand yards away, was the bank of snow which the day before had been piled up against our bulwark.

"Well, captain, it's a quarter of a mile off, anyway," I answered. "Maybe more." From the overhanging wing of the bridge I glanced curiously down on our inclined side, exposed now for the first time in months. Near the water-line, still looking fresh and bright, were those gouges in our elm doubling we had received in early September while butting and ramming a way through that twisting lead into the pack. Looking at those battle scars, I wished fervently that we had had less luck that day in battering our way in. But that was a subject the rights and wrongs of which were now never discussed among the officers. Instead, scanning our listed masts and our unsupported port side, I asked:

"What in the name of all that's holy is keeping us from sliding clear?"

"God knows, I don't," replied De Long solemnly. "I just can't figure it out. When one side of our ice cradle slides away from us without so much as taking with it any splinters from our hull, it makes my theory that our planking's solidly frozen to the ice on our starboard side seem crazy. For why should the ice attach itself so firmly to the planking on one side, and to the other side not at all? It's beyond me, Melville, why we don't slide off." He adjusted the furry edge of the hood of his parka around his eye-glasses, peered down a second at the scarred side below him, then while his glasses were still bright and clear, stared off toward the wall of snow topping the edge of the departed pack, and finally nodded his head as if agreeing with my estimate of its distance.

Looking worn and haggard, for if possible our captain had had even less sleep than any of us during the past week, De Long finished his examination, eyed for a long time his crew stretched out below us along the rail, then turned to me:

"Melville, you're older than I. In the late war you were at sea fighting the rebels when I was still a midshipman, and you've been through lots besides. So I feel I can talk to you, and lean on you as on no one else on this ship, and God above us knows, I need someone here to lean on! Every morning I pray to Him for our safety, every night I give thanks to Him for our escapes during the day. But here in the Arctic, God seems so distant, and this steady strain on my mind is fearful! Look at my men below there, look at my ship! Neither my men nor my ship are secure for a second, and yet I can't take a single step for their security. A crisis may come any moment to bring us face to face with death—and all I can do is to be thankful in the morning that it has not come during the night, and at night that it has not come since the morning! And that's the Arctic exploration I've brought them on! Living over a powder-keg with the fuse lighted, waiting for the explosion, would be a similar mode of existence! Melville, it's hardly bearable." And then looking down again at the crew, he muttered wearily:

"But I've got to keep on bearing it. Call me if anything happens, chief. So long as we're still hanging on here, I'll try to get some sleep now." With sagging shoulders eloquently proclaiming his utter exhaustion, he slumped down the ladder and off the bridge, leaving me alone, figuratively to add an "Amen" to his estimate of our situation.

For over a week, the listing *Jeannette,* which looked as if the pressure of a little finger would send her tumbling out of her inclined bed, nevertheless clung to her half cradle in the pack, defying apparently all the principles of physical force so far as I as an engineer understood them.

On the third day after the pack separated, we had a bad southeast gale blowing all night and all day, with terrific squalls at times reaching a velocity of fifty miles an hour. Although that wind hit us squarely on the starboard bow, its most favourable angle for casting us adrift, the *Jeannette* held grimly to her berth and nothing happened. Then on the fifth day, urged on by a northerly blow, the floebergs again got under way, broke up the young ice to port of us, and jammed themselves under our bows with heavy masses of ice pressing directly on the stem. We confidently looked to see the ship knocked clear this time, but evidently other floebergs jammed against our exposed side exerted such a heavy beam pressure that we stayed in place, though the poor *Jeannette,* squeezed both ahead and abeam, groaned and creaked continuously under the stresses on her strained timbers. The sixth day, the seventh day, and the eighth day, we had more of the same, with streams of floebergs bombarding our exposed port side, and on the starboard side our floe steadily dwindling under the impact of the bergs hurtling through the canal there.

Life on the *Jeannette* became almost impossible. Sleeping with our clothes on, jumping nervously from our bunks at every sudden crackling in the ship's timbers, at each unexpected crash of the bergs outside, we got slight rest for our bodies and none at all for our nerves. And in the middle of all this, the sun disappeared below the horizon for good, leaving us to face what might come in the continuous gloom of the long Arctic night. According to Danenhower's calculations, we could expect the sun to rise again in seventy-one days, unless meanwhile we drifted farther to the northward, in which case, of course, our night would be still further prolonged.

On the ninth day since the separation of the pack, the wind rose once more, blowing directly on our starboard beam, and the never-ceasing stream of bergs began again to pile up across our stem, for us an ominous combination.

On the tenth day, fearing the worst, we rounded up all our dogs, and waited. The pressure ahead increased, with floating ice piling up along the port side higher than our rail, finally starting the planking in our bulging bulwarks. Under the bowsprit, the rising ice blotted from sight our figure-head. Then an up-ended floeberg crashed violently into the pack under our starboard bow and wedged its way relentlessly toward our side. The pressure became tremendous. Beneath our feet the *Jeannette's* tortured ribs groaned dismally. On deck we looked silently at one another, waiting. Something was going to collapse this time. Which would give way first, ship or ice?

Suddenly the *Jeannette* lifted by the stern, shifted a little in her cradle. Instantly the floeberg under our starboard bow drove forward, split our floe, and with a lurch and a heavy roll to port we slid into open water, afloat and undamaged, on an even keel once more!

Intensely relieved at having got clear without being crushed, we nevertheless looked back sadly, as we drifted off among the floe-bergs, at the shattered remnants of the ice cradle which for two and a half months had sheltered us, to see it now tumbling about in elephantine masses, no longer a haven of refuge in our trials.

Well, we were afloat. It was at least some consolation to have a level deck beneath our feet while we waited, sailors with no control whatever over our ship, for what next the ice-pack had in store for us.

But the pack gave us a respite. Idly we drifted about in a wide bay of broken ice, stopping for a brief time alongside one floe, then drifting off till stopped by another. The wind moderated, the temperature rose somewhat till it stood near zero, and finally it began to snow. There being no signs of imminent danger, the captain ordered the bosun to pipe down and we went below, permitted at last to eat a meal without having the plates threaten to slide each instant off the table.

CHAPTER XV

WHAT next?

I had thought that our experiences so far had sufficiently numbed my nerves to enable me to stand anything further with comparative calm, but I had under-estimated the ice-pack.

During the night after our going afloat nothing happened as we drifted, but by early morning of the next day, November 25, we were once more in action, fighting the pack for our existence.

At 6 p.m., as a preliminary, drifting ice pressed us against the edge of the pack and piled high up against our side, nipping our port bow. An hour or so later, this developed into a heavy squeeze which started more of our bulwark planking, and listed us sharply to port. At this, coming while we were at breakfast, things commenced to look bad and we began to shuffle nervously in our chairs, all hands eyeing the exit to the poop, but Danenhower tried to ease the mental strain for us when, bending down to retrieve his spoon which for the third time had rolled off the sloping table to the deck, he remarked:

"Well, mates, if you've been itching for the good old days, now's your chance to cheer. With this heel, the *Jeannette's* beginning to feel to me like the old home again!"

But nobody laughed, and when after an hour of that ticklish heel, the pressure slacked and we levelled off, no one regretted the missing list, unnatural as its absence now felt to us.

Coming up on deck after a hasty breakfast, we found ourselves adrift again near one end of a narrow lead of water perhaps a couple of miles long, at the far end of which appeared a sizeable open bay. De Long debated earnestly with both Chipp and Dunbar, whether he ought to attempt to run out an ice-anchor and make fast to a floe, though the question was largely academic, for never did a large enough floe come near us.

In the late afternoon with still no decision, the question was

settled for us when a strong current springing up for some unknown reason, the rudderless ship began to drift stern first down
that canal in the pack. At the same time the broken ice behind us
also got under way and started to follow, bearing down ominously
for our bows.

We moved along with increasing speed, to our deep relief
steadily gaining on the broken floes pursuing us, till unfortunately
at a bend in the canal, our stern took the bank and stopped us dead.
At this, with our rudder post anchored in the floes, it seemed as if
we were caught, when De Long sang out happily:

"Look! Her bow is paying round as prettily as if she were casting
under jibs!" and to our surprise, it was so. Our stem swung through
a complete arc of 180°, our stern drifted clear of the ice, and there
we were, wholly without effort on our part, properly headed down-
stream with the current!

But even that slight delay while coming about promptly put us in
difficulties. As our stern drifted free of the bank, the oncoming ice
struck us and we were jammed through that canal to an accompaniment of tumbling and shrieking masses of ice awful to contemplate.
Huge hummocks, tons in weight, overhung our bulwarks, threatening to break off and crash down on our decks; floebergs large as
churches bobbed up and down alongside like whales, seemingly
about to come aboard and overwhelm us, time after time leaving
us breathless as huddled inboard round the mainmast we watched,
not daring to go near the rail, even more afraid to seek shelter below.
Helpless, the *Jeannette* was pushed, rammed, squeezed, and hammered along amidst the screeching of the floes. Just as helpless, we
stood in the Arctic night thankful nevertheless for the bright moon-
light which at mid-afternoon was flooding the scene, for had we
without that moon been in darkness forced to stand by and listen
to that shrieking ice without being able to see, God alone knows
what effect terror would have had on us!

This hair-raising passage lasted half an hour. Then as suddenly
as our ordeal had started, it ended in the midst of an eruption of ice-
cakes by our being spewed from a final jam blocking the canal into
a large open bay where the current, with room to spread at last,

quickly lost speed, and the terrifying floebergs, no longer constricted, fell slowly away from our sides!

With fervent sighs of relief at our deliverance we saw the battered *Jeannette* lose headway, float gently toward the wide floe forming the southerly bank of the bay, and quietly ram her blunt nose into the young ice there, by bringing up without a tremor and holding fast. So ended our day.

It was getting along toward the end of November. For three days after that, we lay against the edge of the bay while the young ice thickened about us and a heavy south-east gale kicked up. Our useless masts and spars whipped and rattled in the squalls, our rigging, swollen to two or three times natural size by coatings of frost, sang in the wind in a deep bass pitch wholly new to us, and the ship shook in the gusts as if her sticks were going to be torn clean out of her. But to us as sailors none of this was wholly novel; our only anxiety was what effect this gale, the worst we had yet seen in the Arctic, would have on the pack. We chopped a hole in the young ice alongside, got a lead-line down, and soon observed that the whole pack was drifting to leeward with the wind, moving to the north-west apparently into a large water space temporarily existing unseen by us somewhere there in the Arctic Sea. These drift observations gave us cause for sober thought. What would happen if, with the gale still blowing to urge the ice north-west, something across its path brought the pack to a stand?

We soon found out. On the third day of the storm, in the dim light of a moon just rising in the morning, we saw the leeward ice commence to move past the ship, paradoxically going to windward. Whatever it was, something *had* brought the drifting pack to a sudden halt, but the gale still howled on, driving to the north-west, and unfortunately for us as we lay broadside to it, driving the ice to windward fairly on to our port beam, dead against our framing. We were in for another squeezing by the pack.

Before long the *Jeannette,* with the pressure squarely on her ribs, caught now between opposing floes extending her entire length, was quivering and snapping worse, I think, than ever in our experience. Our spar deck arched up under the strain pitch and oakum were squeezed out of the seams, and a bucket full of water standing on a hatch on the poop was half emptied of its contents by the constant agitation.

To leeward of us, where the ice appeared weaker, one sheet rode up over another, and against this double thickness of ice our starboard side jammed, while the port floe (which for some reason seemed stronger than the ice to leeward) pressed fiercely against us there. The *Jeannette* thus gripped, shivered and groaned dismally and her decks bulged upwards till the heavy athwartship trusses in her hull below came into play and took the squeeze directly. When the ship was able to give no further, the noise ceased, and for half an hour perhaps with only the trembling of the decks to indicate the struggle, the pack pressed and the *Jeannette* resisted while we as helpless spectators waited the outcome.

Suddenly the port floe humped and crumbled, relieving the thrust. Our sprung decks flattened out to normal; we gasped in relief. But our thankfulness was premature as it turned out, for piling its broken edges higher against our side, the port floe, driven in by the wind, pushed up for another nip and the whole performance was immediately repeated with the *Jeannette* in a few minutes as badly squeezed as before.

For eight solid hours the *Jeannette* fought the pack, over a dozen times seemingly compressed to the point of collapse, only to have the floe ice crumple up first and let her spring back into shape each time. There was nothing we could do to aid her—as De Long put it, it was simply a question of the ice going through her or of her being strong enough to stand it. She was strong enough, which was all we could say, and when at last in the late afternoon the gale died down, the pressure ceased and she was still intact, we said it fervently. A good ship, the Arctic Steamer *Jeannette*.

CHAPTER XVI

D ECEMBER, 1879, our fourth month in the pack, came in with crisp cold weather; and as the days passed with the ice about us thickening and the pack showing signs of some stability, we began again to breathe without the subconscious dread that each minute was to be our last. After a few days thus, we even settled into the winter routine of the ship, released our dogs, and commenced to take some interest in the wonders of the Arctic night.

For a month, under the shadow of death, personalities had been forgotten, personal idiosyncrasies submerged. Now with the easing of that strain, our likes and dislikes, our personal vanities, and the ordinary problems of existence in the Arctic, popped up once more.

De Long began to worry over scurvy. No Arctic expedition previously of which we had knowledge had been free of it; in many of them, scurvy, even more than ice, had been responsible for their tale of horrible suffering, death, and disaster. Over-much salt was apparently the cause of scurvy; proper diet, proper water, and proper exercise were the antidotes prescribed by Dr. Ambler, and De Long plunged vigorously into a programme designed to protect us from that loathsome disease.

Exercise to fortify our bodies, the easiest of the requisites to provide, received immediate attention. On December 2, after the first night in weeks during which the captain felt secure enough to take off his clothes when turning in, came a new order.

We were lounging round the mess-room, hungrily waiting for breakfast, while Tong Sing padded about between pantry and table, setting out the oatmeal, the coffee, and the thick slices of bread when the door from the captain's state-room swung back, and with a grave:

"Good morning, gentlemen," in came De Long, holding a paper in his hand.

"Good morning, captain," we replied in a ragged chorus, and

hardly waiting till the skipper had seated himself, slid into our chairs. As usual, I lifted the cover of the oatmeal dish and started to serve.

"Wait a minute with that, Melville; I want to read this order." The captain adjusted his glasses, stroked his moustaches a moment while scanning what he had written, then in his scholarly manner read:

"Until the return of spring, and on each day without exception when the temperature is above thirty degrees below zero, the ship will be cleared regularly by all hands from eleven a.m. till one p.m. During this period every officer and man will leave the ship for exercise on the ice, which should be as vigorous as possible. No one except the officer entering the noon observations in the log will for any purpose during this period return to the ship.

(Signed) GEORGE WASHINGTON DE LONG,
Commanding."

De Long, as he finished, passed the paper to the executive officer on his right, and ordered crisply:

"Chipp, have all the officers initial this now, and then publish it to the crew at quarters." In a more conversational tone, he added to us, "I suppose, gentlemen, the order's obvious enough. We've got to go and get some exercise or we'll all stagnate in this darkness and make it easier for scurvy to get us. I've chosen the time when at least there's a little twilight, even though the sun's gone. Does anybody have any suggestion regarding exercises?"

The paper (together with Chipp's pencil) passed back and forth across the table as one after another, starting with Chipp, we initialled the order, but no one had any comments to make. Once more I started to dish out the oatmeal. Danenhower, at the foot of the table, signing last, tossed the sheet of paper to Tong Sing, who shuffling across the ward-room, with an Oriental bow laid it down before the captain.

"Here, Chipp, take this to read to the crew," said the skipper, starting to push it toward the exec., then on second thought, holding it an instant while his eyes glanced perfunctorily down the column of initials below his signature. A deep flush came over his cheeks as he read and he stiffened a little in his chair, but without looking up, he announced sternly:

"Mr. Collins, I see you failed to sign this. What's the matter?"

There was an instant of tension, then:

"Collins, isn't here yet, captain," put in Chipp swiftly. "He's often late for breakfast. Thinks that having to take the observations on the midwatch is such a strain, he's got to sleep in every morning to recuperate, I guess. I'd tell him later."

"Oh, yes, I forgot that Mr. Collins is not usually with us for breakfast." The skipper's flush faded, he finished pushing the order to Chipp. "Very well, have him sign when he shows up. Now with respect to the exercise for the crew, Chipp, serve out a couple of footballs. They may want to play. And tell them that anyone who wishes can get permission to take a rifle and go hunting."

"Aye, aye, sir!" Chipp folded the order, shoved it into his jacket. "But I'm not so keen on that hunting business, captain. Skulking around through all these broken hummocks, the men'll be shooting each other or the dogs, thinking that they're bears or seals or something. It always happens."

"I won't shed any tears over the dogs, anyway," growled Dunbar. "I think shooting a couple of dozen of 'em 'by mistake' would be a good thing!"

"Belay that, Dunbar, you wouldn't be so heartless," piped up Danenhower. "Don't destroy my last boyhood illusion. What would life in the Arctic be without our dogs, anyway?"

"Still hell, Dan, if you ask me, either with or without 'em," replied the ice-pilot grimly, passing his plate to me for oatmeal. "But getting back to the question of exercise, cap'n, I think letting the men hunt's a fine idea. Surprising how far a man goes thinking that at the *next* water-hole he'll surely get a seal!"

The surgeon laughed softly.

"He'll be surprised all right if he goes with you, Dunbar," drawled Ambler. "I've done it and I know. Every time you say a thing's a mile away across this ice, the only reason it isn't two miles off is because it's three. The men'll be surprised all right if you take them hunting."

Virginian and Yankee, the doctor and the ice-pilot, were off again on their favourite argument, Dunbar's gross under-estimation of the distances he covered on his many scouting trips over the ice. But I had another problem on my mind, and as soon as I had washed down my oatmeal with the hot coffee (which by now Danenhower had managed to get Ah Sam to turn out as a strong black concoction) I went on deck to struggle with my distilling apparatus.

Historically, there is no doubt that scurvy, the seaman's curse since the days of Noah's voyage in the *Ark,* has always resulted on long cruises from the absence of fresh vegetables, the over-abundance of salt beef, and the impure water (contaminated from the bilges) which marked the sailor's diet. And no one who has ever seen the swollen joints, the rotting teeth, the hæmorrhages under the skin, and the bloated faces of the victims, but strains to fight shy of scurvy as a shipmate.

Fresh vegetables, the first defence against this scourge, we could only carry in limited degree when we left San Francisco, and they had long since been exhausted. Of canned vegetables, especially tomatoes, we had a considerable supply and on these we leaned heavily as an antidote. Then of course we had three barrels of lime-juice, the specific remedy introduced in 1795 by Sir Gilbert Blake with such good results in the British Navy that ever since then the British tars, forced to drink the stuff regularly, have been called in derision "limeys" by their Yankee cousins. But in spite of all this we did not feel safe. Other Arctic expeditions within the last fifty years, as strongly fortified as we with lime-juice and in some cases as well supplied with canned vegetables, had before the end of a winter in the ice found scurvy decimating them in spite of their precautions.

We were fitted out with copies of every printed record of polar exploration that either in the United States or in Europe, Bennett or his satellites on the *New York Herald* could lay hands on. And De Long, a good student if the Navy ever produced one, spent hours in his cabin poring over the accounts of his contemporaries and his predecessors in the ice puzzling out that riddle. Why in spite of lime-juice and canned vegetables, in spite of pure fresh water daily replenished from melting ice, had even our immediate rivals in the race to the Pole still fallen prey to scurvy?

Their books gave no answer, but our experiences in getting water by melting ice from the floebergs round us soon gave us a clue. We had been led to believe that when sea water froze under very low temperatures, the salt in it crystallised out, rose to the freezing surface as an efflorescence, and was washed or blown away, leaving the ice free of salt and fit to be melted into good drinking water. Indeed, Dr. Kane, whose words at that time were accepted as gospel truth on all matters Arctic, had written:

"Ice formed at a temperature of $-30°$ Fahrenheit will yield a perfectly pure and potable element."

And confirming this, Lieutenant Weyprecht, of the Austrian expedition which in latitude $81°$ N. had discovered Franz Josef Land only a few years ago, said that they found that ice over "a certain thickness" yielded a pure water.

We were confident therefore when we entered the pack that we needed only to send out a party with pickaxes to obtain from the nearest convenient spot on the floe an abundant supply of fresh water for drinking, cooking, and washing purposes. But we were unpleasantly surprised to discover that we could find not a particle of ice anywhere, whether cut from the top, the bottom, or the middle of the floe, whether taken from old floes fifteen feet thick or young ice a foot thick, that did not contain from twenty to thirty times as much salt per gallon as even the poorest water Dr. Ambler felt he could safely allow our men to drink continuously.

During our initial few weeks in the pack we regarded this situation with incredulity, the same incredulity I have no doubt that the medieval alchemist displayed when his dabbling revealed a fact failing to conform to the principles of matter set forth by the master, Aristotle—it simply could not be so! We concluded at first that perhaps the ice immediately around us had not been formed at low enough temperatures, or that it had not yet had time to reach that "certain thickness." But having nevertheless to get drinking water the while we waited for temperature and time to form round us the pure ice for our permanent supply, we were reduced to scouting far and wide over the floes, scraping together from drifts here and there enough snow to melt up for our minimum needs.

But as 1879 faded into 1880, we drifted to the northward, and the Arctic winter struck us in all its cold fury, we were given a choice opportunity to try to our hearts' content ice of every thickness, formed under every temperature from barely freezing down to −60° F., and we could no longer blink the facts. On this matter, the masters from Dr. Kane to Lieutenant Weyprecht were as reliable as a lot of gabbling old witches—what they said was not so!

In the absence of any startling geographical discoveries or of any marked progress toward the Pole, that we had exploded a third Arctic fallacy (those respecting the Kuro-Si-Wo Current and Wrangel Land being the first two) gave to Captain De Long and Dr. Ambler a sense of having accomplished something at last. For Dr. Ambler deduced from the observed fact that all floe ice retained some salt, the mystery of the scurvy problem in previous expeditions. These, using floe ice more or less mixed either with pure snow or ice formed from melting snow, had obtained water passably potable but actually (though their fixed misconceptions kept them ignorant of it) containing so much salt that in spite of lime-juice rations and what-have-you in the way of canned vegetables, the scurvy had struck them down.

That deduction made it simple for us. All we had to do was to avoid the use of tainted floe ice and we would be the first Arctic expedition in history to dodge the scurvy. And in case the *Jeannette* Expedition discovered nothing else, to bring that discovery back home would at least salve in some measure our pride as explorers.

But if we were not to use the floes, where then was our water to come from? The obvious answer seemed to be from carefully-selected snowdrifts, but as we floated north with the pack, we learned the futility of that. The drifts we relied on for the first weeks after we entered the pack were soon used up and Nature never replenished them. Apparently off the north coast of Siberia in the early autumn it snowed, but as we drifted to the north of Wrangel

Land and the temperature, falling far below zero, stayed there, to
our dismayed astonishment we learned that in the ordinary sense
it never snowed where we were! Apparently the intense cold froze
all the vapour out of the atmosphere, leaving such a trifling percentage in the dry air that regardless of other favourable conditions
for a fine snowfall, there just wasn't enough moisture to provide
the makings. The result was that in a gale when a temperature
change brought snow, all that fell was a fine powdery deposit, ice
mainly, which driven by the wind, cut into our faces like needles.
What was worse for us however (for in most cases we could stay
inboard during a blow) was that the gale drove these particles over
the pack with such force that they acted like a sand-blast on the
surface of the floes, with the net result that when the wind died,
such drifts as we could find were so complete a mixture of powdered
floe and driven snow as to be heavily salted and wholly unfit for
human needs.

Now, while we could find no newly-formed safe drifts, it had
not been wholly impossible for us to get sufficient good snow from
the old ones by going farther and farther afield in the pack until
the last gale in November. This, after making us "shoot the rapids"
so to speak in that canal, had left us stranded miles from our
original refuge in a pack of what was mostly relatively young ice.
Naturally, there were no old drifts in that vicinity and the captain,
at first fearful of being torn away at any minute, was reluctant to
permit anyone to get out of sight of the ship in searching for snow.
Willy-nilly, therefore, we got our water by scraping the tops of
near-by drifts formed in the last storm. This was so salty, however,
that within two days Dr. Ambler had several of the officers and most
of the crew under treatment for diarrhœa. Aside from the ordinary
effects of this disorder in reducing the vitality of those afflicted, to
us it was especially disastrous, for since the "heads" on the ship were
for obvious reasons shut up, we had for months been using portable
"heads" made of tenting, set up on the ice some little distance from
the ship. It needs little imagination therefore to understand what
diarrhœa meant to a man under the frequent necessity of hastily
rushing off through the Arctic night to a flimsy canvas tent to sit

there in the bitter cold of a temperature some thirty degrees or more below zero.

Given a few weeks of such excessive salinity in our water and it was obvious that scurvy would get us, but that at least would take several weeks. De Long was faced with the imperative necessity of rectifying the situation within a few days or of risking the loss of his crew as a result of the unavoidable physical exposure which diarrhœa entailed under our peculiar circumstances.

De Long, Ambler, Chipp, and I held an ambulant council of war. Muffled in our parkas, we first searched the pack around us for suitable snowdrifts in the forlorn hope that perhaps the men had missed a good one. We found a few that to the taste seemed passable, but in each case the hope faded when the surgeon squeezed a drop of silver nitrate into a melted sample, and inevitably the milky white reaction showed excessive salt.

Not very hopefully we scanned the "head" situation. No chance of improvement there. Since the ship was immovably frozen into the ice, we dared neither to re-open the "heads" on the ship nor bring the ones on the floe any closer to the gangway without risking an outbreak of contagion.

So there being no safe water available from the pack ice, no hope of getting any from snowfalls, and the absolute need of providing some quickly lest the next movement of the ice find us with a helpless crew unable even to abandon ship, it was the conclusion of the council, that regardless of cost, we must make our own from sea water. Naturally enough, since I was engineer officer, De Long turned that problem over to me.

Ordinarily it would not have been much of a problem technically. On the ship steaming normally, and feeding her boilers from the sea, I might have bled some steam off the auxiliary line, put it through a distilling coil or worm we had fitted in our engine-room, and collected the resulting fresh water. But we were not only not steaming normally, we were not steaming at all, because for the reasons I have given previously our fires were out, our fire-rooms were cold, and our boilers were emptied.

Aside from that, there was another angle to it that griped the

I

captain. To take sea water and distil it over into fresh water you've got to boil it. That takes heat, and heat takes coal, and coal was of all things we had aboard the most precious, more so even than food, for in a pinch with our food exhausted we might go out on the pack with rifles and knock down bears, seals, and walruses enough to exist on, but where in those icy wastes could we go to knock down even one ton of coal to feed our boilers when our bunkers were emptied? For we had left only ninety tons, which (save for the scanty supply I doled out to Ah Sam daily for cooking, and to Bosun Cole for stoking the two stoves forward and aft to keep men and officers from freezing to death) under the captain's orders I was religiously husbanding, so that if ever we were released by the pack, we might be able again to fire up our boilers and do some of that exploring for which we had come north.

Up to now, to live at all, we had had to burn coal enough to run the galley and our heating stoves; from now on, if we were to live without scurvy, we would have in addition to burn coal enough to run some kind of an evaporator. What kind it might be, to give us safe water and still consume the least possible quantity of "black diamonds," the captain left to me.

The problem started not with "How much water do we need?" but with "How little water can we get by on?" I canvassed this question with the doctor, the captain, the exec, Ah Sam and finally Jack Cole—all of whom had something to contribute on what was the least possible quantity needed for drinking, for cooking, for tea, and for washing—and I came out with the answer that forty gallons of water a day, about a gallon and a quarter for each one of our thirty-three men, was the irreducible minimum.

Naturally for this quantity, which was more or less in line with the daily capacity of any really ambitious Kentucky moonshiner's still, it was foolishness to think of firing up so large a kettle as one of our main boilers. Thinking over what else we had, my recollection lighted on a small Baxter boiler which we had brought along to furnish steam for driving an Edison electro-magnetic generator and illuminating the ship with his new-fangled carbon lamps. Edison's generator having proved a flat failure (probably because it got soaked in salt water on our stormy crossing of Behring Sea) the captain had ordered the whole works dismantled and struck below into the hold. Without further delay, I had Lee and Bartlett resurrect the Baxter boiler (leaving the rest of the outfit below) and this little boiler with the help of my machinist and fireman, I soon had rigged up inside the deck-house, with its steam outlet hooked to a small coil set outside in the open air on top of the deck-house, where the cold air would act as a very effective condenser on the vapour passing through the worm.

Meanwhile, not waiting for this contraption to get into action, at the surgeon's suggestion the skipper ordered Cole to break out from the hold a couple of barrels of lime-juice, which on December 2 for the first time on the cruise, he started to issue. In our mess, a pitcher of this stuff was placed on the table at dinner, where under the watchful eye of the surgeon, each one of us, sweetening it to taste, had to drink an ounce. For the crew, Alfred Sweetman, carpenter, was given the responsibility of seeing that the men took theirs, and as each watch laid below for dinner, under Sweetman's observation, each man was handed a tin cup with his ration of lime-juice and an ounce of sugar to sweeten the unsavoury mess, and compelled to drink it before he could draw his food ration. Months of storage in casks had not improved its flavour any, so in spite of Ambler's gaze and Sweetman's vigilance, had it not been for the sugar generously served out to sweeten the dose, I have little doubt that, scurvy or no scurvy, all sorts of ingenious dodges would shortly have been developed to avoid swallowing that tart medicine.

When the last pipe-joint was tightened up, Bartlett fired the Baxter boiler and we commenced distilling. Our first few days at it were to my surprise pretty much of a failure, for the distilled water which we collected up on deck in a barrel set underneath the outlet of the condensing worm, while better than the melted snow, still tested far too high in salt for safe use, and our diarrhœa continued unabated. This puzzled me (not to mention severely disappointing the captain) and it took some hours of sleuthing about to discover the trouble. I then found that we were feeding the boiler from a tank atop the deck-house. This tank was filled by the seaman on watch, who hauled water to the topside in a bucket from a hole chopped in the floe alongside. Unless the man was careful (and a sailor working outside in a temperature of 30° below zero is interested only in speed and not in care) he would slop the sea water over both coil and deck-house, from which places enough trickled down into the fresh-water barrel to ruin completely our day's output. Having discovered this, I promptly rigged a pan over the barrel to catch the drip and looked hopefully for better water. But my hopes were dashed once again when, watching Surgeon Ambler test a sample from our next barrel of water (the result of a whole day's distilling), I saw to my disgust the sample turn as milky as ever immediately he dropped a little silver nitrate into it.

By now, we had been suffering four days from diarrhœa and the situation was serious. I dropped everything else to devote my whole time to watching the operation of our evaporator, endeavouring by an analysis of what I could see done and what theoretically must be going on inside the apparatus from fire-box to receiving barrel, to locate the reason or reasons why from our sea-water feed, we failed to get over and condense a pure steam, leaving all the salt behind as a brine in the boiler. Thinking at first we might be boiling off the water too fast, I had Bartlett damp his fire somewhat to make less steam, but I soon found that that solved nothing. For with too little steam going up through our condensing coil in the frigid atmosphere outside, the condenser promptly froze up and burst a pipe, putting a stop to distilling altogether till Lee thawed out the coil and repaired the leak.

But hardly had we resumed operation again when what I saw gave me the answer. Bartlett started up his little feed pump, and began vigorously to pump cold water into the hot boiler to bring up the level in the glass. Promptly, as shown by the needle on the gauge, the pressure in the boiler tumbled and the water in the sight glass started to bubble vigorously. And that had been our difficulty. The sudden injection of cold feed water evidently created a vacuum in the steam space. Under the reduced pressure the hot water in the boiler had boiled off so violently that it carried salt spray up with the steam and over into the distiller, where it ruined our make.

Now I had it.

"Enough, brother!" I sang out to Bartlett. "Stop that pump, haul fires and secure everything!"

And from then on, alternating between sweating over that hot boiler and freezing on our enforced trips to the "head," Bartlett, Lee, and I struggled all through the night. We shifted the location of the feed water line inside the little boiler to a point as far away from the steam space as we could get it, and inserted a constriction in the steam line to the feed pump so that no one could, even by accident, start the pump suddenly or make it stroke at anything more than dead slow speed.

In the early morning we finished, refilled the boiler, fired up and again started distilling. When we had to feed the boiler, we fed slowly (which was the only way the pump would now run) and I felt sure from the slight fluctuation of the pressure gauge that I had at last ensured operation steady enough to eliminate priming. And when at noon with the barrel half-full of distilled water, Bartlett, Lee, and I, in the front row of a cluster of fellow sufferers, gathered wearily round the surgeon as he poised his silver nitrate solution over the test cup, I felt there was some warrant for the hearty cheer which echoed down the deck when Ambler announced:

"Very pure, chief!"

So ended our struggle to get fresh water. And in a few days our intestinal troubles ended too, a result for which all hands were devoutly thankful. But when I reported our success to the captain, while he was even more laudatory in congratulating me than anyone else had been, still for him there was a fly in the ointment which completely took the edge off his enthusiasm.

"How much coal does that distiller use up, Melville?" he queried.

"About two pounds of coal per gallon of water made, sir," I answered.

He figured mentally a moment, blinked sadly at me through his glasses, then muttered:

"Two pounds per gallon, chief? Why, it's nearly a hundred pounds of coal a day just for distilling! That expenditure will ruin us if we have to keep it up. Snow, snow! That's what we need!"

CHAPTER XVII

CONTINUING his programme for dodging scurvy, De Long followed up his exercise order by another calling for a thorough monthly medical examination of all hands. In this I believe he had two objects—the main one, of course, to give the surgeon a chance to catch and deal with the first symptom of disease and especially scurvy, before it had any opportunity to get out of bounds; the other, by maintaining a record at frequent intervals of our physical condition, to study the effect of the long Arctic night and of Arctic conditions generally on the human body, and to learn perhaps the best method of combating these effects.

I read the order absent-mindedly, made a mental note that at ten next morning I was due for examination, and in the midst of my engrossment over the urgent problem of how to save some coal, promptly put the matter out of my thoughts. An hour later, Charley Tong Sing touched my shoulder and announced in a sing-song voice:

"Captain wantee you, chief, in cabin allee samee light away."

More discussions about coal economy, I presumed.

But that idea was quickly knocked out of my head when stepping into the cabin I found myself facing not the captain alone, but also Mr. Collins.

A little surprised at this unexpected situation, I looked enquiringly from one to the other. Both men were on their feet, both were angry, and evidently trouble was in the offing. Not being invited to take a seat, naturally I remained standing also, looking quizzically from Collins to the captain, wondering what was up.

I found out soon enough. De Long, waiting only till he was sure that the steward was out of the room and the door firmly closed behind me, with an evident effort to maintain an even tone, broke the silence.

"Melville, I've sent for you as the officer aboard with longest experience in Service customs to get from you an independent opinion on the propriety of my medical examination order before I proceed to enforce it. It seems that Mr. Collins here objects."

So that was it.

I swore inwardly. Here was Collins heading for trouble again, and unfortunately for me, here I was dragged into the muddle, evidently by the captain this time, and from the nature of the case, bound to offend our meteorologist if I even opened my mouth. What ailed Collins anyway? I had never seen a man on shipboard with such an unholy penchant for getting himself into difficulties.

Apparently the wrinkling of my bald brow and the way I fingered my beard as the situation hit me, gave Collins an inkling of my feelings, for without giving me a chance to speak, he burst out heatedly:

"You're absolutely correct, I do object! And regardless of what Melville or anybody else you bring in here may say, I'm going to keep on objecting! I never liked that exercise order you've already issued, even though I'm obeying it. I'm a grown man, and I was before ever I saw this ship, and I've got sense enough to decide for myself how much exercise I need to keep my health and when I need to take it, without anybody telling me. I don't need to be ordered out like a schoolboy for supervised play, nor have my steps dogged like a poor man's cur to see I take it. Nevertheless, I swallowed that. But this is too much! I've got some rights and I've got some pride! Even if I am down on the shipping articles as a seaman, I'm not a damned guinea-pig, to be stripped naked every few weeks for the doctor to experiment on!"

This time I guess my jaw did drop in open-mouthed astonishment. That *seaman* business again. How it must be rankling in Collins's soul! I looked from Collins's overwrought face to De Long's, flushing a fiery red. Had he been any other skipper I had ever sailed with, I should have seen Collins immediately clapped into the brig for gross insubordination. But of the scholarly De Long's reactions I was not so certain. Prudently I closed my mouth without uttering a word. There was nothing I could say anyway that wouldn't make a bad situation worse.

De Long's blue eyes, a startling contrast to his burning cheeks, blinked queerly through his glasses as he stood there, struggling inwardly to control himself the while regarding Collins.

"The scholar in him's going to win out over the sailor," I thought to myself. "There'll be no arrest."

And so it proved. For what seemed an oppressive length of time under that strain, the captain, without speaking, glared at Collins and Collins unflinchingly glared back. Finally, in an unbelievably mild tone, the captain broke the tension.

"Will you please be seated, chief? I should have asked you before." I sat down. "And that will do for you, Mr. Collins; you may go now. I see that Mr. Melville and I will get along much more rapidly discussing this subject without your presence."

Like an animal suddenly uncaged Collins, still glaring, turned his back on us, and broke from the cabin, leaving the door wide open. The captain closed it, then sank into an arm-chair facing me, nervously chewing the twisted ends of his moustaches. Still breathing heavily from his repressed emotions, he turned to me:

"Melville, it seems that everything I do for the discipline and safety of my crew, that man takes as a personal affront! And now over this examination matter, he's positively insubordinate! I sent for you that we might discuss that order in a reasonable manner, and find out what's wrong with it, if anything is. But you saw what happened instead! Nevertheless, chief, I want your frank opinion. Is there anything wrong with that order?" De Long paused, looked anxiously at me.

"To tell you the truth, captain," I said, "I read it only once hurriedly and then never gave it a second thought. The Navy Regulations require us all to stand an annual physical examination; what difference it makes to anyone, except to the doctor who has to do the work, if it's monthly, I can't see. But so long as Dr. Ambler isn't complaining, what's Collins blowing up about it for?"

De Long shook his head wearily.

"I don't know, unless he can't get it out of his head that I'm persecuting him. That hallucination of his about being a seamen started him off on it long ago. Congress wrote the law commissioning the *Jeannette* under which he shipped—I didn't. He had to ship that way or not at all, but Heaven knows I've treated him as an officer in spite of it! A lot of good it's done. I try to make every allowance for his point of view, but there is a limit. I can't let him defy me on this medical examination. Even if I were so derelict in my duty as to allow discipline to be flouted by such mutinous conduct, I just can't take chances on having a sick crew in our desperate situation!"

"Right enough, captain," I agreed. "I should think even Collins would see that. He's an intelligent, educated man. But I think there's something in addition to the persecution bug that's biting him this time. Did you catch the inflection he put on that word 'naked'?"

"I'm afraid I was so astounded at his words, I missed his inflections," confessed the skipper. "What about it? What's wrong with 'naked' here, inflected or not? There's not a woman within a thousand miles of us to embarrass anybody."

In spite of the gravity of the situation, I grinned inwardly at that.

"Well, captain," I said, "so much the worse for us. I just have an idea that's one reason this crew's all so glum. But that's not what I was aiming at in Collins's case. Women don't enter into his ideas of embarrassment. It's all in the way he was brought up. He's a sensitive person, almost morbid, I'd say, and the idea of having to strip before anybody, especially under what he thinks is compulsion, gripes his ideas of dignity and personal privacy. Now, I'm not excusing insubordination, sir, but with Collins's peculiar civilian background in this expedition, since you've asked for it, I'd suggest a modification of that order that'll still get the results and not hurt anybody's feelings. Of course the change can't be for him alone; that would never do—but why not modify it so's the doctor examines all the officers stripped to the waist only, and all the crew stripped completely? That'll have two good effects. It won't require anything of Collins that offends his dignity, and it'll show him that he's getting better treatment than the 'seamen' he's so wrought up about being classed with. Then if anything's ever going to clear the cobwebs out of his brain and stop his belly-aching, that'll do it."

To De Long, already overburdened with a sense of failure and the weight of the Arctic problems menacing us, and sincerely desirous of maintaining harmony amongst his personnel, this appealed as a sensible solution. He nodded approvingly.

"A good idea," he agreed, expansively relaxing in his chair. "I'll do it! And much obliged to you for the suggestion, chief. It helps a lot to feel I can always rely on you to lend a hand when there's anything wrong, whether with the machinery or the men."

"Hey, brother!" I cautioned, "easy on taking in so much longitude in your thanks. Better wait till you see how it works. I'll guarantee the machinery on this ship, but God himself won't guarantee the men!" and with that I took my departure and returned to my evaporator, leaving the captain to re-draft his order for the medical examinations.

To a degree, it worked. Collins, who it seems had submitted a written protest in addition to expressing himself so freely orally, when he read the revised order asked leave to withdraw his objection, and submitted himself (though very sullenly) to the examination which Dr. Ambler carried out in the privacy of his cabin. And the captain, who, boiling under Collins's insolence, had been ready to hang him for it, calmed down despite the fact that in a measure Collins had won, and accepted the situation, treating Collins as courteously as if nothing had ever arisen.

Collins, however, not appreciating his luck, failed to reciprocate. Ever since the bear-hunt incident, he had refused to ask the captain's permission to go on the ice, staying aboard except when his routine observatory duties (and now the enforced exercise order) gave him the opportunity to leave the ship without asking. Instead, he had ostentatiously paced the deck, indulging in what he was pleased to inform us was "a silent protest," which obviously gave him great satisfaction, though why I don't know, for De Long diplomatically took the sting out of that performance by totally ignoring it. Now Collins withdrew still further into his shell, avoiding the captain altogether except when duty made it impossible, and what was worse for him, taking to avoiding the rest of us also when he conveniently could, a proceeding which hardly added to the sociability of the ward-room mess. He even refused to say "Good morning" to any of us when first we greeted him in the mess-room, and this boorishness soon put him completely beyond the pale of our little society.

Queerly enough, Collins now began associating almost exclusively with the very seamen with whom he took such violent objection to being classed, spending most of his time with my fireman Bartlett, and retailing to him and thus to the crew generally, practically every bit of ward-room gossip that he heard. Such a situation was hardly desirable aboard ship, and De Long endeavoured to put an end to it by privately conveying to our meteorologist the information that such association was decidedly contrary to naval custom and that it was beneath his dignity as an officer so to consort with enlisted men. But the captain's friendly admonition only drew more black looks from Collins, leaving De Long more perplexed than ever over Collins, who refused to comport himself either as officer or seaman, and leaving Collins with his persecution mania flaring up even more fiercely.

December dragged along. The ice around us kept freezing thicker and thicker under the intense cold. On the surface, the pack held together, but despite that, kept us uneasy. Night and day (by the clock, that is, for so far as light went, it was always night for us except for a semi-twilight around noon) even in calm weather we were likely to be disturbed by noises like the beating of the paddle-wheels of innumerable steamers and by occasional terrifying shocks on our hull, all of which kept us jumpy. At first we had no explanation for this uncanny state of affairs, the pack around us showing no movement and the ship being solidly enough frozen in.

But Dunbar finally solved it for us. As he pointed it out, evidently we were now suffering from a bombardment of underrunning floes. Considerable masses of ice thrust under the pack in the November break-ups were kept constantly in motion by the current beneath the re-frozen surface. They bumped along as best they could under its ragged contour, giving that paddle-wheel effect, and naturally enough, when one collided with our submerged hull, giving us the unpleasant sensation of having struck a rock.

An understanding of the situation, while removing the mystery, did not greatly help our peace of mind. None too sure in the light of our past experiences, of the solidity of the newly-frozen pack, we were for ever standing by for an emergency with sledges, boats, knapsacks, and provisions ready to go over the side. The monotony of continually expecting trouble with none of the excitement of actually seeing things happening, had its own peculiar effect on us, making sound sleep impossible, killing our appetites, and leaving us restless, listless, and haggard, a condition which the severe physical discomforts of our situation naturally aggravated.

Still, for all our nervousness, we began to note some strange things, the results of the intense cold which descended on us. The atmosphere, practically free of moisture, was startlingly clear, and never have I seen such brilliant stars as shone down on us from those December Arctic skies. Then (owing perhaps to the increased density of the cold air) sounds on the ice travelled unusual distances and boomed and reverberated as if from an overhead dome or the roof of a mammoth cave. And the auroras, shimmering across the sky in a dance of vivid colours, were indescribably beautiful. But what struck us most, around thirty degrees below zero, was the almost unbelievable effect of the cold on the ice itself. Subjected to a temperature far below its freezing point, the ice assumed a flinty hardness and strength entirely different from its normal state. The floes grating against each other, instead of crumbling under pressure, gave out an unearthly high-pitched screech. And when we went out with picks or axes to dig away the ice in the fire-hole under our stern, granite itself could not have been more effective than that cold ice in turning the edges and blunting the points of our tools.

Finally, there was another effect of the extremely low temperature which most of all racked our nerves. Standing, sitting, or sleeping, who can accustom himself to having pistols unexpectedly discharged practically in his ears? Yet we were constantly exposed to such nervous shocks. For all over the ship, the iron fastenings of our planking and our timbers, contracting abnormally from temperatures never expected by the builders, compressed the wood under the bolt-heads as the iron shrank till the wood, finally able to stand no more, suddenly snapped with a noise like a pistol-shot. And so startling was each such explosion in one's ears, so like a pistol discharge, that even the thousandth time it happened, involuntarily I jumped as badly as the first time I ever heard it.

Even the poor dogs suffered unexpected trials and I well believe that to their canine souls, their difficulties were quite as trying as ours. Like Dunbar, I had little natural sympathy with the vicious brutes and saw little value in their presence, but having been to some degree a party to transporting them from their usual habitat, I could not but feel some responsibility for their new troubles. And queerly enough it fell to my lot as engineer partly to relieve them.

Aneguin and Alexey, our two Indians, were primarily responsible for our forty dogs. Each day in the forenoon watch, they fed them, bringing up from the forehold from the cargo of dried fish we had taken aboard at Unalaska, the necessary amount for issue, one dried fish per dog per day being the authorised ration. Ordinarily, the wise dogs immediately crushed their fish in their powerful jaws and swallowed them in one gulp; the otherwise dogs (a pun I fear almost worthy of Collins) found themselves fighting for the remains of their fish with their mates who were quicker on the swallow, a habit which always made feeding time alongside ship a bedlam. Without particularly paying any attention as to why, I noted vaguely that as December drew on, this daily snarling of the dogs over their food subsided. As a minor blessing I was duly grateful, until one day coming aboard a little late after my prescribed exercise period, I saw Alexey on the quarterdeck performing an autopsy on a dog which following a brief illness the afternoon before had died during the night. As I approached, Alexey removed from the dog's stomach a wad of oakum as big as a baseball, the very evident cause of his death. I squeezed the ball, incredulous. Oakum, all right. But why should even an Eskimo dog eat that? I asked Alexey. Between pantomime and Indian English he explained it to me:

"Fish in hold freeze, chief. Verr hard. Dog chew. Verr hard. Lak iron. No good chew." He seized a marline-spike, went through the motions of a dog trying to chew a fish frozen presumably as hard as iron, and very plainly breaking his teeth on it. He laid down the marline-spike. 'No good. No chew fish, no swallow. Dog get ongry. Bym bye eat oakum. Bym bye die." Sadly he waved at the deceased dog.

That explained the cessation of our daily dog fights at feeding time. The fish stowed in our hold had frozen so hard there that no dog, no matter how energetically he chewed, was now able to masticate his own fish quickly and get it down. As a consequence, all the dogs being in the same boat, too busily engaged trying to chew their own dinners to bother about stealing each other's, there were no fights. But this poor devil, his teeth apparently unable to make any impression on the fish, had been driven in desperation to something softer and had unwittingly committed suicide by gobbling the oakum.

I grunted sympathetically. A dog's life, all right. But I could fix it. Motioning Alexey to follow me inside the deck-house, I had him bring up from the hold one day's issue of fish, only thirty-nine now. They were frozen hard, no question; even with a crow-bar, it would take a strong man to make a visible impression on one of those glaciated fish. Sizing up their approximate volume, I had Lee make a sheet-iron box large enough to hold the lot, and fit inside it a few turns of pipe which I connected to the blowdown from our evaporator, the Baxter boiler. Alexey tossed in the frozen fish, and Lee put on the cover.

"That'll thaw 'em out, Alexey," I informed him. "Every time we blow down the hot brine from that boiler, it'll heat the fish, and in a few hours, they'll be soft, even a dog with false teeth won't have any trouble with 'em. Now don't forget; fill the box every night, and by morning dinner for the dogs will be all ready."

Alexey, a very good Indian and deeply concerned for the well-being of his charges, thanked me profused, and judging by the resumption of the snarling over dinner next day, I guessed the dogs had reason to also.

But the dogs had still one more cross to bear that I could not ease. Their instinctive habit in cold weather was to bed themselves down at night in soft snow, keeping themselves as comfortable that way as an Eskimo inside his igloo of ice. But if we had reason to regret the absence of snow because it deprived us of a source of fresh water, the dogs lamented its absence even more because it robbed them of their natural beds. Night after night they wandered round the ship disconsolately looking for drifts, and finding none, were forced at last to turn in on the bare ice. For some time we had noticed each morning here and there hair embedded in the ice, but when the December cold snap hit us, we were surprised to find several dogs with so much hair frozen to the ice that they just could not tear themselves free. There was, however, nothing we could do about that except to make it Aneguin's regular detail to go out before feeding time each morning with a shovel and break out from the floe all the dogs that had been frozen down the night before, a job which required great finesse with the shovel on Aneguin's part, lest all our dogs soon become as bald as Mexican hairless poodles.

CHAPTER XVIII

MONOTONOUSLY the dreary days drifted by. In darkness we ate our food, took our exercise, thawed out our frozen noses afterward, and vaguely wished we could "go somewheres." December 22, the shortest day of the year came, bringing with it, aside from the most brilliant display of auroras we had yet witnessed, only the knowledge that with the sun at its extreme southern declination, half of our seventy-one-day long night was gone. But the day itself was further marked by the fact that Mr. Dunbar, that veteran whaler and the only member of our mess who had ever before wintered inside either the Arctic or the Antarctic Circles, came down with a bad cold. His tough hide had according to his own claim always before resisted illness, so this made him doubly miserable, and he moped around the ward-room very low in spirit. Finally, as if to make sure that we remembered the day, Danenhower also complained that his left eye pained him, and after a session with the doctor, big Dan completed our picture of wardroom woe by coming in with a black patch over the ailing optic, explaining that Ambler had found it somewhat inflamed and had advised him to give it a rest by shielding it even from the poor glow of our oil-lamps for several days.

Two days later we came to Christmas Eve, which for us, except for plenty of ice around, was everything that traditionally Christmas Eve is not. No children about, eagerly excited over hanging up their stockings; no friends dropping in; no families, no wives, no sweethearts—nothing of these for any of us, but instead only the memories of bygone Christmases under happier circumstances, and the hope (clouded by gnawing doubts) that another Christmas might see us out of the ice and restored home.

We gathered in the ward-room, a glum group—Dunbar nursing his cold, Danenhower with his black patch looking like a pirate in distress, Ambler, De Long, Chipp, Newcomb, and myself. Only

Collins was missing. That his presence would have added any gaiety was questionable, but that he saw fit to stay locked in his state-room, keeping the ward-room bulkhead between himself and us, certainly added to the general gloom. And gloomy it certainly was in that room—a smoky oil-lamp the only illumination, the warped wood panels of the bulkheads the only decoration, overhead the deck beams heavily covered with insulating layers of felt and canvas, dismally sagging under the weight of the combination of frost and moisture with which they were saturated, and beneath our feet the sloping wood deck, wet from the condensate dripping off the cold forward bulkhead.

I did the best I could to lighten matters up. Back at the Mare Island Navy Yard before we left, Paymaster Cochran had thoughtfully presented me with a bottle of fine old Irish whisky, which I had so far carefully hoarded. Now I broke it out from beneath my berth, scraped together some other ingredients, and with all hands watching, mixed a punch in the soup tureen. In the damp chill of the barren ward-room, we filled our glasses, lifted them.

"To Cochran!" I proposed. "May he yet be Paymaster-General!"

With no disagreement to this, we all downed Cochran's whisky, and warmed a little by the fiery Irish spirits, promptly refilled our glasses. There was just enough for a second round. I looked questioningly at De Long for him to propose the second and (of necessity) last toast. Whom would he choose, James Gordon Bennett, the sponsor of our venture; the President; someone more personal, perhaps?

But Danenhower gave him no chance. Lifting his glass, he waved it over the empty bowl, swept us all with his one uncovered eye, and sang out:

"To our old shipmates, Emma De Long and Sylvie—may they never have cause to worry over us!"

That also I could heartily endorse, so wasting no regrets over the amenities due Bennett or the President, I raised my glass to drink as did the others, when Dunbar alongside me poked me in the ribs. I leaned over toward him.

K

"Mrs. De Long's all right with me," he whispered, his voice hoarse from his cold, "but who's this Sylvie?"

"Captain De Long's daughter," I hissed. "You old fool! Drink it down before he knocks you down!"

"Oh, all right," mumbled Dunbar. "I thought maybe she might be Newcomb's sweetheart." He drank his whisky at a gulp.

And that just about ended our party. With no more punch to serve as an excuse for conviviality, the conversation soon faded into the general murk gripping the room, and with everyone seemingly immersed in memories of happier Christmas Eves, one by one all hands drifted away to warm over their recollections in the solitude of their state-rooms.

Christmas Day, mainly because it lasted longer, was even more dreary than Christmas Eve. A high wind and biting clouds of fine snow made going on deck or on the ice wholly uninviting. Confined again to the ward-room or to our state-rooms, we moped over our memories, tried to imagine how friends, relatives, or families were spending the day, and thought a little enviously of Navy shipmates in port the world over with vessels decorated from deck to trucks with wreaths and garlands of greenery, and ward-rooms echoing with the alluring laughter of women troubled with no deeper problem than how after dinner to get a husband or a sweetheart excused from watch and off the ship.

We did have a grand dinner, to provide which Ah Sam performed miracles with the humdrum materials available in the store-room, topping off all with mince-pies soaked in brandy. The eating of this unexpected banquet almost made us forget our surroundings and our situation. But not quite, for we ate our dinner to the constant rumbling of the unseen pack, the occasional explosive snapping of timber fastenings, and even a few sharp shocks from underrunning floes. And like a death's head at the feast, to show that all was not joy and brotherly love on the *Jeannette* on this Christmas Day, there next to Danenhower at the foot of the table was Collins's chair—empty, while Collins, sulking in his state-room, dined alone.

I think I misstate nothing when I say that in the ward-room of the *Jeannette* we were all thoroughly grateful to see the last of that

Christmas Day, and I have little doubt that each of us fervently prayed ever to be spared another like it.

December dragged away. We came to the end of the year 1879. To help the crew in welcoming in the year 1880, on which now he banked heavily for success, the captain sent forward four quarts of brandy, while I did what I could with a fifth quart to provide good cheer for the ward-room mess. As a result, when the rapid ringing of the ship's bell at midnight marked the birth of 1880, the whole crew (despite a temperature nearly 40° below zero) gathered on the quarter-deck just outside De Long's cabin, gave three cheers for the *Jeannette*, sent an embassy of two into the ward-room to wish us all a Happy New Year, and then hastily beat a retreat to the berth-deck to warm up on those four bottles.

This evidently so heartened the crew that after their New Year's dinner (mince-pies and brandy once more) they staged an entertainment, the high-lights of which were Aneguin imitating Ah Sam singing over his kettles, and a prompt and contemptuous imitation by Ah Sam of an Indian attempting to imitate a Chinaman, which performance brought down the house.

This comic relief for a brief while took our thoughts off what our more sober senses looked forward to with misgivings in contemplating 1880. Under our noses, so to speak, as we emerged from the crew's entertainment to the deck, was the unpleasant discovery that the mercury in our thermometers had frozen at −40° F., unobtrusively suggesting thereby that what we had so far seen of Arctic temperatures was merely an introduction to what was yet to come.

A second more disquieting situation was that Danenhower's eye inflammation had grown worse. The doctor had that day been forced to put him on the sick-list, confining him to his room in absolute darkness because the slightest light falling on his eye caused severe pain. Aside from the fact that the loss of his services threw an added load on the remaining officers—the captain, Chipp, and myself—in carrying on the ship's work, his condition gave us real cause for worry. In case the ship went out from under us, leaving us stranded on the ice, there was a blinded and a helpless officer on our hands to care for, probably requiring to be dragged every inch on a sled, for it was as much as even a man with two perfectly good eyes could do to get over that rough pack without breaking his neck every few steps.

What had caused Danenhower's eye troubles? All of us, from the first day we were caught in the pack until the sun in November vanished for good, had religiously worn snow-goggles, for the glare off the ice was intolerable to face. Why had Danenhower, the youngest regular officer we had and physically by far the most powerful member of the ward-room mess, been knocked out by eye failure when neither forward nor aft had anybody else in the ship's company been so much affected? Puzzling over that, I could conclude only that it was an unfortunate combination of his job and his personal characteristics functioning under very unfavourable circumstances. Dan was navigator. Innumerable times he stood under terrible conditions of cold, straining his eyes through his sextant, trying to get with poor horizons (or with an artificial mercury horizon) shots at the sun, the moon, or the stars to establish our position as we drifted with the pack. That was bad enough, but what apparently was worse was that Dan was the most painstaking and the most indefatigable worker over account books I ever saw aboard ship. In addition to being navigator, he was our supply officer, and hour after hour he had pored over coal reports and store-room records, figuring and re-figuring, trying to keep track of and account for each pound of coal used, almost each ounce of flour expended. Under the poor lamplight by which since early November he had worked continuously, the load on his eyes, already overstrained by constant squinting through sextant telescopes, proved too much and an inflammation enveloped his left eye, shortly developing into an abscess which threatened to blind that eye completely and even involve the other one. The result was that in a desperate effort to save his sight, the doctor was forced to make Dan a prisoner, forbidden (except when completely blindfolded, he was led out for meals) to leave the darkness of his room. And few prisoners in history, regardless of the horrors of their medieval dungeons, ever had a worse outlook to face than Danenhower in his pitch-black cell—small, damp, chilly, and with always the rumbling and screeching of the pack to remind him that any day the unseen walls of his prison might collapse and the prison itself sink from under him, leaving him helpless on the ice.

Over all of us, his shipmates, Danenhower's disaster threw a pall of gloom that New Year's Day. Over De Long, who felt a special responsibility for each man in the ship's company, it fell like a blight, evoking apparitions for 1880 of calamities yet undreamed of.

So ended our holiday season—a dismal Christmas and a worse New Year's, leaving us with the temperature starting downward from −40° F. to face whatever new the pack had to offer.

January drew along, bringing gales, biting clouds of flying ice particles, and deeper cold. The ice, getting denser and denser as it grew colder, shrank, and about the middle of the month cracked open, forming little canals on both sides, leaving us in a small island of ice hardly a ship-length across. We contemplated that dubiously, for if any pressure came from the pack about us, now forty inches thick, we would receive almost directly the thrust of the pressing floes with no protection at all. But luckily no pressure came before the extraordinary cold rushed to our rescue by freezing the water which welled up in the fissures. The first half of the month, therefore, went by with only the usual monotonous groaning and rumbling of the pack and occasional nips on our hull to keep us in mind of our position.

January 19, 1880, was on the other hand, a red-letter day for us. In the silence following the subsidence of a gale which was in no way worse than many another we had experienced, for no reason apparent to us the floe into which we were frozen began early in the morning to crack and split in every direction. Promptly the anchor watch sent word below, and as usual, we all came tumbling up on deck, there to remain stock-still in our tracks as awe-struck we watched in the unearthly half twilight of the Arctic a sight entirely new to us.

North, south, east, or west, it was the same. In a large circle surrounding the ship, the surface of the pack was everywhere heaving up into a ring of rugged mountains high above the level of the sea! Huge masses of ice, large as ocean liners, pitched and rolled on the crests, while reverberating from all about came a shrieking and a screeching from the tumbling ice that froze the very marrow in our bones. Like the jaws of a slowly closing vice, that circle drew in on us—ahead, astern, on either beam—whichever way we looked there was an approaching mountain of ice steadily, relentlessly advancing on the *Jeannette* across the small expanse of yet unbroken pack, while on that undulating ring with cracks streaking across it like forked lightning, the floes parted with roars like thunder, forming a deep bass background for the "high scream" of the flint-like ice of grating floebergs, the whole echoing across the pack to us in a veritable devil's symphony of hideous sounds.

The ring was still a quarter of a mile away.

On the bridge, Captain De Long, eyeing it, cupped his hands to try to make himself heard above the din, bellowed to those on the spar deck below:

"All hands! Stations for abandon ship!"

Listlessly we moved to our stations abreast the loaded sledges on the poop, but what could we do? Enclosed on all sides by that shrinking circle of tumbling ice, where could we go for safety when we abandoned ship? Even unencumbered by sledges or knapsacks there was not a chance in the world of scaling the slopes of those moving mountains of ice against the stream of floebergs cascading down their sides. Flight was impossible, annihilation certain!

Dunbar was by my side. With a sea-going eye he scanned the little plain of unbroken pack still surrounding us, then muttered:

"That ice is approaching us at the rate of a fathom a minute. It's still sixty fathoms off. In sixty minutes, chief, we'll all pass over to the Great Beyond!"

Apparently he was right. Motionless, silent for the most part, we stood, clinging to our useless sledge-loads of pemmican. That terrifying ring, irresistible, inexorable, shrank in on us. Numbly we waited for that avalanche of ice to come tumbling aboard, crushing us like flies, crushing our ship.

On it came. Fifty fathoms, forty fathoms, thirty fathoms. Then as inexplicably as the motion had started, it stopped, a ship-length or so away, leaving us after an hour of looking death squarely in the face, limp, completely drained of emotions, and incredulous almost of being still alive. Slowly the hills of ice flattened out, there remaining around us an indescribable "Bad Lands" of broken floes; and the shrieking died away into a strange quiet except for the rumble of under-running floes bumping along in the current beneath our pack. It was over, we were safe, our vessel undamaged. Yet had the ship been a few hundred feet in any direction from the exact spot in which she lay, she would inevitably have been lost, and we with her.

Feeling like men reprieved when the noose had been tightened about our throats and the trap all but sprung, we left the poop slowly, noticing for the first time how cold we were. But in spite of that, curious to examine at still closer range the danger we had so narrowly escaped, all hands except those on watch clambered down the starboard gangway to the ice and were soon dispersed among the nearest slopes, climbing the pinnacles, gazing in awe at some of the nearer floebergs standing up-ended from the pack, and speculating on the results had this or that colossal berg capsized on our vessel.

On the *Jeannette,* five bells struck. It was 10.30 a.m., the time for serving out the daily allotment of coal for the galley, the heating stoves fore and aft, and the distillers. Regretfully I turned my back on the marvellous vista of ice-peaks and canyons stretching before me, and with a frozen nose, bleary eyes, and a beard white with frost from my heavy breathing, started stiffly back to the ship. I wanted to make sure that young Sharvell, the most inexperienced of my four coal-heavers, whose turn it was to break out the coal from the bunkers, was not imposed upon either by the guile of that Chinaman, Ah Sam, or the bullying of the bosun into passing out a pound more of our precious fuel than was allotted them by my orders.

I climbed the gangway, crossed the deck to the machinery hatch, and was half-way down the ice-covered iron ladder, just turning on the middle grating to descend into the fire-room, when in that darkness, as if the devil were after him, a man came bounding up the ladder, rammed me in the stomach, and nearly ricocheted me off the grating to the fire-room floor below. I saved myself only by grabbing his arm as he shot by. But to my surprise, instead of stopping, he struggled to tear loose and continue on his way. In the gloom, I peered at him. It was little Sharvell, my coal-heaver, apparently badly frightened, his rolling eyeballs and pallid face startlingly white against the smudges of coal-dust on his forehead. Well, I was no doubt as white, having just had my wind completely knocked out by his carelessness, and I was mad besides. I tightened my grip on him.

"You clumsy cow," I gasped, "wait a minute there! What d'ye mean by——"

The next I knew, I was talking to myself. Sharvell, twisting free, was racing up the ladder.

Thoroughly enraged now, I shouted after him:

"Damn you! Come back here!"

But Sharvell did not come back, he kept on climbing. The thought, however, of coming back penetrated his fright enough to loosen his tongue, for he yelled down to me:

"On deck, quick, chief, while you got a chance! The ship's sinking! The fire-room's flooded already!"

CHAPTER XIX

THE *Jeannette* sinking? Sharvell must be crazy. The ship had gone through far worse squeezes before without a leak. Nevertheless, forgetting our encounter, I raced down the ladder. There was the fire-room entirely flooded from port to starboard, with water already over the floor-plates and rising steadily toward my empty boilers!

For a second I stared in cold dismay. No steam on the ship to run a pump. If the water rose over our furnaces before we got our fires going and steam up, there would never be any steam—and we were through! Once the water got that high, nothing under Heaven could prevent the ship from filling at her leisure and sinking from under us. I had to get steam and get it fast!

On deck, Sharvell had already spread the alarm. Even as I watched the rising water, sizing up my procedure, estimating my chances of getting steam before the water got us, overhead I heard the noise of running feet, guns being fired to recall the men on the ice, the shrill piping of the bosun, and Jack Cole's stentorian call:

"Man the pumps!"

Man the pumps? Why man them? There in the engine-room a little abaft me, their bases already in the water, were my frost-covered pumps. What they imperatively needed in those frozen cylinders was steam, not manning. Then it came to me. The *Jeannette* was a sailing ship as well as a steamer—she still carried hand-pumps, the same crude hand-pumps with which Columbus had kept his leaky caravels afloat. And they might save us too; keep the water down below the level of my furnace grates till I raised steam!

And now came action. Down the ladder to join me slid my black gang—Lee, machinist; Bartlett, fireman; all my assortment of coal-heavers—Boyd; Lauterbach, the German; Iversen, the Swede; and even that frightened little Englishman, Sharvell. In the biting

cold of the fire-room, 29° below zero (it was 45° below outside),
I hastily detailed them.

"Lee! Get aft into the engine-room and line up the main steam-
pump to suck on this fire-room as soon as you get steam!"

"Bartlett! Outboard there with you! Open the port sea-cock and
flood the port boiler to the steaming level. Open her wide, and
four bells on that flooding!"

"Lauterbach! Get some kindling wood down here from the
galley! Shake it up now, and mind you keep that kindling dry!
And while you're on the topside, tell the skipper I'm firing up the
port boiler!"

"Iversen, you and Sharvell start breaking out the coal. Get plenty,
and keep it in the buckets, out of this water!"

"Boyd! Spread the fuel in both furnaces in the port boiler as
fast as it comes to you, and get an oil torch going, ready to light
off when the water's up to level!"

In the faint gleam of a few oil-lamps in that frigid fire-room, off
the men splashed through the ice water on the floor-plates, an in-
congruous group for a black gang if ever there was one, as clad all
in furs from hoods to boots they stumbled away to their stations in
a temperature more suitable to the inside of a refrigerator than to a
boiler-room.

On deck, I heard the clatter of equipment and the banging of
mauls, Cole's shouts, the hoarse responses of running seamen, and
the curses of Nindemann and Sweetman struggling to break out
crossbars and handles frozen to the bulkheads and rig the hand
gear for working the forward bilge-pump—a tough job in that sub-
zero atmosphere on the topside with everything iron shrunk by the
cold, everything wood swelled by frost and moisture, and nothing
fitting together properly as it should.

But long before they got the hand-pump on deck assembled, I ran
into troubles of my own. Bartlett, wrestling with the port sea-cock
(I had chosen that side because the ship being heeled to starboard, it
was the only one still showing above water), his stocky frame and
brawny shoulders straining against the wrench, sang out to me:

"This cock's frozen, chief! I can't get her open!"

I jumped to his aid. Together we heaved on an extension handle to the valve wrench. No movement. I was desperate. We had to get that cock open to the sea or we could not fill our boiler. More beef was needed on the wrench. I looked inboard. There in the dull light of the oil torch in his hand, before the port boiler waiting for fuel to arrive, was big Boyd, doing nothing.

"Boyd! Lend a hand here!"

Boyd shoved his torch into the cold furnace, splashed over to us. The three of us, fireman, coal-heaver, engineer, braced ourselves against a floor stringer, put our backs into it, heaved with all our might against that wrench handle. The cock gave way suddenly, twisted open. I sighed thankfully, let go the wrench.

"Watch her now, Bartlett," I cautioned. "Wide open to the sea till the water shows half-way in the boiler sight glass, then shut off! Careful now; it'll only take a few minutes. Don't overfill her!"

But I might have spared both my thanks and my caution. Bartlett waited a moment for the water to rush through from the sea into the empty boiler, then feeling no vibration in the pipe to indicate flow, stooped, pressed his ear near (but not too near) the frost-coated sea-cock, and listened carefully. Not a murmur of running water. Bartlett lifted his head.

"No water coming through, chief."

No water? I felt sick. Then that long disused sea-chest must be plugged with ice! Frozen solid where beyond the valve it passed through our thick wood side to the sea, totally beyond our reach for thawing out, effectively blocking off any flow of water. We could not fill our boiler!

I cursed inwardly. Literally we were sunk now. Caught with no steam, boilers empty, unable to get water into them to raise steam, what good to us now was all the coal we had saved for our exploring by that economy? There the saved coal lay, worse than useless in the bunkers, serving only to ballast down the ship that she might sink the faster under us!

Thump, thump! Thump, thump!

From on deck came a welcome sound. The carpenters had at last got the handles rigged. The hand-pump was starting! With four men on each side swaying over the bars, vigorously putting their backs into each stroke, that steady thumping gave me new hope. If the hand-pump, inefficient though it was, could only keep the leak from gaining too fast on us, I still had a chance! Water to fill the boiler? Why bother about the sea? We were standing in an ocean of salt water right there in our fire-room and more was coming in all the time! All I needed was time enough to get a boiler full of it off the submerged floor-plates into that port kettle, and I could light off!

"Bartlett, forget that sea-cock! On top of that port boiler with you and your wrench. Open up the man-hole there, then stand by the opening to receive water in buckets! Boyd, get Sharvell and Iversen out of the bunkers, get some buckets, and form a line to pass water up to Bartlett as soon as he gets that man-hole open!"

Bartlett scrambled over the furnace fronts and up on top of the boiler. Boyd passed up his torch to illuminate the work, and I tossed up a sledge-hammer to help him start the bolts on the man-head. While Bartlett laboured over the bolts and Boyd and the other coal-heavers scurried through the engine-room and the fire-room collecting all the buckets, I stood a moment before the port boiler, sizing up the situation.

Where was all that water coming from, anyway? There was no sign of damage, no sign of leak in the machinery spaces. From forward, probably; we had got some very bad raps on the bow during the morning's excitement. Perhaps an under-running floe had rammed our stem, opened up our forepeak. If such were the case, that hand-pump running on deck forward was in the best location to hold down the water, to keep it from rising too rapidly here amidships. I listened an instant to the rapid *thump, thump* of the oscillating handles, then caught mixed with the noise a husky cry from the men at the pump:

"SPELL O!"

There was a break in the rhythmic thumping, a new gang stepped in and relieved the men at the handles, then the monotonous throbbing was resumed. SPELL O, the cry for relief, already coming from the first gang manning the pumps! Back-breaking work that, all right. How long could the sixteen men we had on deck, even relieving each other frequently, keep those handles flying up and down fast enough to give us a chance in the fire-room? Not for long could human muscles stand that pace, I feared. It would be nip and tuck between us and the rising water.

From atop the boiler came the banging of metal on metal and the muffled curses of Bartlett as sprawled out in the scanty space between boiler and deck beams overhead, he fought with sledge and wrench to loosen the man-hole bolts. Lauterbach came cautiously down the fire-room ladder, balancing a huge armful of kindling. I motioned him to toss it on to the grates, then to join Boyd, Iversen, and Sharvell with the buckets. In silence, we waited below, listening to the mingled chorus of the banging sledge-hammer, the rasping screech of rusty nuts, and the fluent profanity of Bartlett, prone on his stomach, a fantastic fur-clad demon with his distorted face showing up intermittently in the flickering flame of the torch, battling the boiler beneath him. No one could help him; there wasn't room for two men to work in those confined quarters. And there was no use giving him any advice either. So below we stood, straining our eyes impatiently toward Bartlett, while inch by inch the water rose on us and the margin between water level and furnace grates shrank. The hand-pumps on deck were losing out—they had slowed up the rise, but they could not stop it.

My chilled legs felt cramped. Instinctively, not taking my gaze off Bartlett, I tried to flex my knees to relieve them, shifting my weight from one foot to the other. I found I could not lift either leg. Looking down sharply, I saw for the first time what before in the poor light had escaped my notice—in that intense cold, far below zero, the water was turning to slush, ice was forming here and there over its surface, and both my feet were solidly frozen down to the iron floor-plates on which I stood!

I gripped my legs one at a time with both hands, savagely tore them free.

"Keep moving, boys!" I warned the men in the water alongside me. "If you stand still a minute, you'll be frozen down!" And standing there in that fast-freezing water, at 29° below zero, I was at least thankful for the four pairs of wool socks, the three suits of blue flannel underwear, and the two pairs of woollen mittens which encased me under my fur suit and boots, for otherwise by now, between cold water and cold air, I should have frozen stiff as a board.

Bang!

With a final blow of his sledge, Bartlett knocked free the last dog, lifted out the boiler man-head, shouted:

"All clear, chief!"

"Start those buckets!" I ordered, but it was unnecessary. Already Boyd had dipped the first one full, was passing it up to Bartlett, who dashed the contents through the open man-hole into the boiler, where splashing over the frigid iron plates inside, I haven't the slightest doubt but that it promptly became ice.

Round and round went the buckets, Lauterbach filling, Boyd and Iversen passing them up full to Bartlett, and little Sharvell catching the empties as they came tumbling down the boiler front. All the men were soon coated from head to foot with ice from the water slopping from the buckets—only their constant stooping, rising, and twisting which kept cracking the ice off in sheets prevented their soon accumulating so heavy a weight of it as no man could even stagger under.

Meanwhile, as they laboured, I turned to, and took Boyd's place in spreading fuel on the grates, preparatory to lighting off. Hastily I scattered the kindling over the cold furnace bars, then slid several buckets of coal out the nearest bunker door, carefully manœuvring them through the slush and ice across the flooded floor-plates to avoid slopping the sea water which reached nearly to the tops of the buckets, in on the coal. Seizing then a shovel, I started to heave coal into the furnaces, an awkward job, for getting the shovel into the tops of the upright buckets was difficult, and naturally I dared not dump the coal out on the floor-plates first. As best I could, I managed it, spreading the coal over the kindling, a little thin at the front of the grates, a thicker bed at the rear. That done, I leaned back on my shovel, and alternated between watching the water-line creeping up the boiler fronts and my men frantically passing up buckets to fill the boiler.

It was a big boiler, eight feet in diameter, and would require innumerable buckets. Mentally I calculated it, making a rough estimate. Nine tons of water had to be man-handled up into that boiler to fill it properly, a thousand bucketfuls at the very least. I timed the heavy buckets; about six a minute were going up, but the men could hardly maintain that pace. Still, even if they could, it would take three hours to fill that boiler to the steaming level! Long before then, the fire-boxes at the bottom of the boiler would be flooded, we could never light off! Somehow, we had to keep the water down in the fire-room till I got steam, or the *Jeannette* was doomed. And her going meant a two-hundred-mile retreat over the broken pack to Siberia—in mid-January at 40° or worse below zero, an absolutely hopeless journey!

"Keep 'em flying, boys!" I called out to my coal-heavers, "while I lay up on deck for help. I'll be back here in a minute!"

Coated with ice to the waist, I clambered up the ladder, went forward into the deck-house. Swinging on the pump bars there, were eight straining seamen; against the bulkhead, resting a moment, were eight more, including even the Chinamen Ah Sam and Tong Sing. A little forward of them was De Long, anxiously peering down a hatch into the forepeak, while below him in that gloomy hole, Lieutenant Chipp and Nindemann were sloshing round in deep water with a lantern, searching for the source of our troubles.

"Where's the leak, captain?" I asked, bending down alongside him.

De Long straightened up, intensely worried.

"We don't know, chief; Chipp can't find it. All he can see is that the water's gushing through that supposedly solid pine packing the Navy Yard filled our bow with, as if it were a sieve. The leak's in the stem, down somewhere near the keel; I think our forefoot's twisted off." He looked at me with haggard eyes. "We're still holding our own on the forepeak with the hand pump; but the men'll break down before long. How soon can you give us steam and help out, chief?"

I drew him aside, a little away from that squad of resting seamen, not wishing to discourage them.

"Never, captain!" I whispered hoarsely, "unless we get help our-selves!" Briefly I outlined our desperate position. There was no hand-pump in the fire-room, the water was gaining on us there also. "I've got to have a gang to hoist water out of that fire-room by hand someway to keep it down till my boiler's filled and I get steam up, or we're done for! And it'll take three hours yet. My gang's all busy. Who can you spare?"

De Long gazed at me sombrely.

"Except Danenhower, who's blind, every man and officer's work-ing now. But Newcomb and Collins are only collecting records in case we abandon ship. Will they do?"

I laughed bitterly.

"Newcomb isn't worth a damn for real work, captain; and from what I've heard from Collins, you could shoot him before he'd turn to as a seaman! Besides, two are not enough, anyway. It'll take six good men at least to keep ahead of that water, and then they may not do it. But give me Cole and half of that relief gang at the pumps there and I'll try."

"That'll reduce us here to six men a shift on the pump-handles," muttered the captain, dubiously eyeing the crew at the pump. "But we've got to get steam! All right, Melville, take them. But for God's sake, hurry it up!"

"Aye, aye, sir!" I turned abruptly to our Irish bosun, who was near-by supervising the pumping. "Jack, pick four men out of the gang here, any four, and come aft with me. Shake a leg, now!" I started for the after door in the deck-house.

Cole grabbed Starr, a Russian and physically the strongest sea-man in our crew, off the starboard pump-handle; took Manson, a burly Swede, off the port handle to even things up, and beckoned to Ah Sam and Tong Sing from the relief gang.

"C'mon, me byes; lay aft wid yez!" Cole marshalled his little detail out of the compartment and slammed the deck-house door behind them almost before the twelve startled men left at the pump could realise that they now had the work of all sixteen to carry on.

Close outside the deck-house stood the barrel which received the fresh water condensed in our distiller. That barrel was just what I needed; distilling for the present was the least of my worries.

"Jack," I explained briefly, "the fire-room's flooding on us. We got to keep that water down till I get fires started. Sling that barrel in a bridle, rig on a whip to the davit over the machinery hatch, and start hoisting water out of the fire-room, four bells and a jingle! She's all yours now, Jack! Get going!"

Cole, a rattling good bosun if I ever saw one, needed nothing further.

"Aye, aye, sor. Lave ut to Jack!" In a moment he had that Russian, the Swede, and the two Chinamen round the barrel, emptying it; in another second they were rolling it aft; and as I started down the ladder to the fire-room, Cole had the barrel on end again and already was expertly throwing a couple of half hitches in a manila line round it to serve as a sling.

Almost before I got down the ladder to my fire-room again, the barrel came tumbling down the hatch at the end of a fall and landed alongside me with a splash, while above, Cole roared out:

"Below there! She's all yours! Fill 'er up!"

Being nearest, I tipped the barrel sidewise in the water, pushed it down till it submerged, then righted it. It filled with a gurgle, settled through the slush to the floor-plates.

"Full up!" I shouted. "Take it away!"

"Aye, aye!" The line to the barrel tautened, then started slowly to rise. Down the hatch floated Cole's voice encouraging his squad on the hoisting-line:

"Lay back wid yez, Rooshian! Heave on it, ye Swede! An' git those pigtails flyin' in the breeze, ye two Chinks, or we'll all be knockin' soon at the Pearly Gates, an' fer sailor min the likes of us, wid damned little chanct to get past St. Peter! Lively wid yez; all togither now. Heave!"

The loaded barrel suddenly shot up the hatch.

Hurriedly Cole swung it over to the low side scuppers, dumped it, and sent it clattering down again. Once more I filled it, started it up, then called Lee, my machinist, from the engine-room pump to stand by on that filling job while I went back to the all-important boiler.

L

Why go into the agony of the next two hours? Wearily, without relief, my men heaved water, ice, slush, whatever the flying buckets scooped up, indiscriminately into the yawning void inside that boiler; just as wearily, with aching shoulders, Cole and his little group laboured, unrelieved and unshielded from the bitter cold on deck, heaving that barrel up and down; while from the deck-house, the more and more frequent cries of SPELL O! showed that at the undermanned pump, backs were fast giving way under that inhuman strain.

And in spite of all, I could see that we were going to lose. Another hour yet to fill the boiler to the steaming level, but from the rate with which the flood-waters were still rising, in another hour it would be too late—the water would be over the grates. Hoping against hope that perhaps I was wrong, that perhaps the water was going into that kettle faster than I thought, I crawled myself to the top of the boiler. Keeping as clear of Bartlett as the scant space allowed, not to slow up the stream of buckets, I seized the torch and in between the dumping of those cumbersome buckets peered through the ice-rimmed man-hole into that Scotch boiler. As I feared. The upper tubes down there were still uncovered; the crown sheets of the furnaces were still perhaps a foot above the level of the slush (I could hardly call it water) line. As I looked, Bartlett, sprawled out beside me, sent another bucketful splashing through the man-hole, which soaked my beard and almost immediately froze it into a solid mass. But I hardly noticed it, staring with leaden eyes into that still half-empty boiler. With a sinking heart, I slid away on the ice-coated cylinder from the man-hole, and crawled down the breechings to stand once again on the thickening ice covering the flooded floor-plates.

Dare I fire up without waiting further?

I was in a terrible predicament. To light fires under a partly filled boiler like that, with tubes and furnace plates not wholly covered with water, was not only the surest way to a court-martial which would probably end my naval career, it violated also every tenet in my engineer's code, violated every principle of safety, practically insured a boiler explosion! But if I did not get fires going right away,

I would never have a chance to fire up, and not only that boiler, but the ship herself and all her crew besides would vanish in that Arctic ice.

I must risk whatever came.

With flying buckets and tumbling barrel splashing and spilling water all around me, I applied a match to another oil torch, fanned it a moment in the chilly air till it blazed brightly, shoved it (in the narrow space still remaining between the flood waters and the grate bars) into the inboard furnace under the kindling, till the wood took fire and then hurriedly transferred it to the outboard furnace until that also lighted off. The extreme cold of the outside air favoured me, creating a tremendous draught as soon as a little warm air filled the flues, and in no time at all it seemed, the wood was blazing up fiercely and igniting the coal which, shining brightly down through the grate bars on to the water flooding the lower part of the ash-pits, cast a lurid red glare out into the dark fire-room, evidently putting new life into the drooping sailors, for both below and on deck, a ragged cheer greeted that crimson glow.

"Keep that water going, lads; we haven't won yet!" I warned, flinging open the furnace doors and heaving in more coal. "We've got to get that water level up over the crown sheets before they get red-hot, or we're all going straight to hell! Twice as fast now on those buckets!" And whatever it was, fear or hope, that inspired those coal-heavers, a moment before ready to drop from utter exhaustion, the buckets started to fly faster than ever.

I finished heaving coal, slammed to the fire doors, and leaned back on my shovel. I was in for it now. Never in the history of steam, before nor since, has a boiler been fired under such weird conditions —furnaces half-flooded, no water showing in the sight glasses, slush and ice for what charge there was, and the boiler man-head still off! But I was relying on some of those very dangers to save my bacon— till I put the man-hole cover back, there could be no pressure to cause real trouble; and till we had melted down and warmed up that ice and slush, I counted on that chilly mixture and the water still splashing in to soak up heat so rapidly as to keep the bare tubes and exposed crown sheets from getting red-hot and collapsing.

My other fears I need hardly go into—the dangers of bringing up steam suddenly in a cold boiler instead of gradually warming up first for twelve hours as was usual; of frozen gauge glasses; of frozen feed-pumps—all these I deliberately put out of my mind. Only one thing counted now—to get some steam at any cost whatever before the water reached the grate bars and flooded out my fires.

And we did. With only a few inches left to go, came at last from Bartlett the long-awaited cry:

"The crown sheet's covered now, chief!"

"On with that man-head!" I roared back.

The clanking of Bartlett's sledge-hammer, breaking away the ice round the man-hole so the cover would fit, was my only answer. The worn-out coal-heavers dropped their buckets, rested for the first time in hours, sagging back against the boiler fronts to keep from dropping into the icy water. No time for that. I seized a slice-bar, started savagely to slice the fire in the outboard furnace, sang out:

"Boyd, get busy with another slice-bar on that inboard fire! Lauterbach, relieve Lee on filling that barrel! Lee, get back to your pump now! And, Sharvell, you and Iversen, get into those bunkers and break out some more coal! Come to life now, all of you!"

Boyd, nearly dead from his half of heaving up over eight tons of water, staggered over to my side, gripped a slice-bar. Together we laboured over the fires, forcing them to the limit, nursing in more coal without deadening the blaze, till helped by an amazing draught from the stack, we had them roaring like the very flames of hell itself. Never have I seen such fires!

Leaving the stoking job now wholly to Boyd, I dropped my slice-bar and stepped back to examine the gauge-glasses. Water was barely showing in the sight-glass, but, thank God, it was showing! And the needle of the pressure-gauge was starting to flutter off the zero pin. Steam was coming up! If we could only hold down the flood for a few minutes more now, till I could get that pump warmed up and going, we were saved! But that part was up to Jack Cole.

"Jack!" I shouted up the hatch. "A little more and you can quit. But right now, for God's sake, shake it up; faster with that barrel!"

"Aye, aye, sor!" Then to his strangely conglomerate crew, ready undoubtedly to collapse in their tracks, Cole called gruffly:

"C'mon me byes! Lit's raylly git to liftin' now, an' work up a sweat, or we'll freeze to death in this cowld! Lay back on ut, Starr! Heave there, Manson! Wud yez have thim two Chinks outpullin' yez? An' step out there now, ye Chinese sea-cooks, an' don't be clutterin' up the decks, or whin that Rooshian gits goin', he'll be treadin' heavy on thim pigtails! Yo heave! Up wid ut!" And with astonishing speed I saw the loaded barrel vanish up the hatch.

I breasted my way through the water aft to where Lee in the engine-room stood by my largest steam-pump. No need to worry about priming the pump for suction; another foot higher on that flood and we would have to go diving to reach the pump-valves. I felt the steam-line. The frosty chill was gone; a little steam at least was already coming through to the pump.

"All right, Lee; let's get going," I mumbled. We cracked open the steam-valve a hair, started to drain the line. And no mother nursing her baby ever handled it more tenderly than Lee and I nursed that frozen pump, gradually draining and warming the steam-cylinder, lest the sudden application of heat should crack into pieces that abnormally cold cast iron, and after our heart-breaking struggle with the boiler, leave us still helpless to eject the sea. With one eye on Jack Cole's rapidly moving barrel and the other on that narrowing margin between flood water and furnace fires, I nursed the pump along by feel, taking as long to warm it up as I dared without swamping those flames. At long last the pump-cylinder was hot; steam instead of water was blowing out the drains. And the boiler-gauge needle stood at thirty pounds. Enough; we could go.

I straightened up, motioned to Lee to start the pump. He opened the throttle valve. With a wheeze and a groan the water-piston broke free in its cylinder, the nearly submerged pump commenced to stroke.

Leaving Lee at the pump, I ran (that is, if barely dragging one ice-weighted foot after another can be called running) up the ladder toward the deck. While I climbed, the empty barrel came hurtling down the hatchway, splashed into the water in the fire-room. Before Lauterbach could fill and up-end it, down on top of the barrel in a maze of coils came the slack end of the hoisting-line. Apparently Cole's gang was through.

As I poked my head above the hatch into the open, there—— Oh, gorgeous sight for bleary eyes and aching muscles! was a heavy stream of water pulsing into the scuppers! Near-by, prone on the deck where they had dropped in their tracks when they let go the hoisting-line, were four utterly worn-out seamen, gazing nevertheless admiringly on that beautiful stream. And leaning against the bulwark watching it was Jack Cole, who, as he saw me, sang out:

"Praises be, chief; we're saved! There'll be no calls for SPELL O from that chap!"

CHAPTER XX

OUR immediate battle was won, but the war thus opened that 19th of January, 1880, between us and the Arctic Sea for the *Jeannette* dragged along with varying fortunes till the last day I ever saw her.

Our big steam-pump made short work of all the water in the fire-room that was still water. In an hour the room was bare down to ice-coated floors and bilges, with the pump easily keeping ahead of the leakage coming from forward. But the men at the hand-pump, optimistically knocked off the minute the steam-pump began stroking, were unfortunately not wholly relieved. Despite the fact that we opened wide the gates in the forepeak and the forehold bulkheads to let the water run freely aft to the fire-room pump, the flow through was sluggish, impeded, I suppose, by having to filter through the coal in the cross bunker. So fifteen minutes out of every hour, the hand-pump was manned again to keep down the water level in the forehold, while, sad to contemplate, our weary seamen, between spells at the pump, had to labour in the forehold store-rooms breaking out provisions (much of which were already water-soaked) and sending them up into the deck-house to save our food from complete ruin.

It was ten-thirty in the morning when the leak was discovered; it was three p.m. when I finally got steam up and a pump going; but at midnight the whole crew was still at work handling stores. The state we were then in was deplorable beyond description.

Who struck eight bells that night I do not know, for since morning we had had no anchor watch, but someone, Dunbar perhaps, whose sea-going habits were hard to repress, snatched a moment from his task and manned the lanyard. At any rate, as the clear strokes of the bronze bell rang out on that frost-bitten night, De Long, in water up to his knees in the forehold, was recalled to the passage of time. The provisions actually in the water had been broken out; his effort now was to send up all the remainder which rising water might menace. But with the bell echoing in his ears, the captain, looking at the jaded seamen about him, staggering through the water laden with heavy boxes and casks, toiling like mules, came suddenly to the realisation that they had only the limited endurance of men and called a halt.

"Knock off, lads," he said kindly. "If anything more gets wet before morning, it gets wet. Lay up on deck!" And on deck, as the men straggled up the hatch to join the rest of the crew round the hand-pump (at the moment unmanned) he ordered Cole to serve out all around two ounces of brandy each. Frozen hands poured it into chilled throats, to be downed eagerly at a gulp—there was not a man who might not have swallowed a whole quart just as eagerly, and probably then still have felt but little warmth in his congealed veins.

At the captain's order, Cole then piped down—the starboard watch to lay below to their bunks, the port watch for whom there was to be no immediate rest, to man the hand-pump as necessary through the remainder of that dreary night, keeping the water in the fore-hold down below the level of the as yet unshifted stores. The frozen seamen tramped wearily off, some to rest if they could, the others to bend their backs over the bars of the pump, which soon resumed its melancholy clanking.

But neither for me, for the captain, nor for Chipp was there any rest. Immediately I had downed my share of the brandy, I turned to at once, figuring how I might get steam and a steam-pump forward to suck directly on the forehold and eliminate altogether the toil over the hand-pump which must soon break our men down. I had in my engine-room that spare No. 4 Sewell and Cameron pump (which my men and I had so thoughtfully picked up in the dark of the moon at Mare Island before we started). I set to work on a lay-out for installing it in the deck-house forward; which task, between designing foundations and sketching out suction and steam lines for it, kept me up the rest of the night. As for Chipp, he was down in the forepeak with Nindemann, endeavouring to stop, or at least to reduce, the leak. The water was pouring in through the innumerable joints in that mass of heavy pine timbers, which stretching from side to side and from keel to berth deck in our bow, filled it for a distance of ten feet abaft the stem. However valuable that pine packing may have been in stiffening our bow for ramming ice, it was now our curse, very effectively preventing us from caulking whatever was sprung in the stem itself. All through the night Nindemann and Chipp laboured, stuffing oakum and tallow into the joints of that packing where the jets of water squirted through. It was discouraging work. As fast as their numbed fingers rammed a wad of oakum into a leaking joint and stopped the flow there, water spurted from the joints above. Methodically through the night they worked in that dismal hole with freezing water spraying out over them, following up the leaks, caulking joint after joint, but when at last they got to the top, plugging oakum into the final crack, the water rose still higher and started to pour down their necks from between the ceiling and the deck beams overhead where they could not get to it. They could do no more. At 5 a.m., each man a mass of ice, they came up, beaten.

Meanwhile, De Long, forseeing the possibility of such a contingency, had himself put in the rest of the night over the ship's plans, designing a water-tight bulkhead to be built in the forepeak just abaft that packing, so that if we could not stop the leak, we could at least confine the flooding to a small space forward and thus stop all pumping, either by hand or steam.

In the early morning, after twenty-four hours of continuous strain and toil, the three of us met again in the deck-house, I with my sketches for the pump installation, De Long with his bulkhead plans, and Chipp with the bad news that we had better get both jobs under way at once, for he had failed utterly to stop the leak. So we turned to.

I will not go into what we went through the week following— my struggles with frozen lines, improper equipment, and lack of men and tools for such a job. Suffice it to say that after three days I got that auxiliary Sewell pump running forward, so that to the intense relief of the deck force, their torture at the hand-pumps ended altogether, and I was able to keep the water in the forepeak so low that Sweetman and Nindemann were enabled to start building the bulkhead.

From then on, Nindemann and Sweetman bore the brunt. On these two petty officers, Sweetman, our regular carpenter, and Nindemann, our quartermaster (but almost as good as a carpenter) fell the entire labour of building that bulkhead. In the narrow triangular space in the peak, they toiled hour after hour, day after day, cutting, fitting, and erecting the planking. William Nindemann, a stocky, thickset German, was a perfect horse for work, apparently able to stand anything; but Alfred Sweetman, a tall, spare Englishman, had so little flesh on his ribs that he froze through rather rapidly, and in spite of his objections, had to be dragged up frequently to be thawed out or he would soon have broken down completely. As it was, every four hours both men got a stiff drink of whisky to keep them limbered up, and as much hot coffee and food in between as they could swallow, which was considerable.

Meanwhile, during all this turmoil and anxiety, the captain was weighed down with the problem of what to do with the blinded Danenhower should the water get away from us, either then or later. To add to his worries, Dunbar, who was also still under the weather from his illness, seemed between that and his efforts to assist, to have aged overnight at least twenty years. It was pathetic to see the old man, looking now positively decrepit, struggling in spite of the captain's orders to hold up his end alongside husky seamen, fighting with them to help save the ship. And as if to make a complete job of De Long's mental anguish during that agonising first day of the leak, Surgeon Ambler was suddenly taken violently ill, and to the captain's great alarm had to be left in his cabin, practically unattended. Aside from De Long's natural concern over what might happen to Ambler himself, the effect on the captain's mind of this prospect of being left without a doctor to look after Danenhower and any others who might collapse in our desperate predicament, can well be imagined. It amazed me that the captain under the combined impact of all these worries and disasters, instead of caving in himself, maintained at least before the men an indomitable appearance, by his actions encouraging them, and with never a word of profanity, urging and cheering them on.

By the end of the ensuing week things showed signs of improvement—I had both steam-pumps going, hand-pumping was discontinued. Nindemann and Sweetman against terrible odds were making progress on the bulkhead, Dunbar was no worse, and Ambler (whose trouble turned out to be his liver) was under his own care, sufficiently on the mend to be no longer in danger.

Only Danenhower, aside from our leak, remained as a problem. He, instead of getting better, got worse.

The third day of our troubles, while I was still struggling with a frozen steam whistle line through which I was trying to get steam forward to start my Sewell pump, there into the glacial deck-house beside me came our surgeon, wan and pinched and hardly able to drag one foot after another. I gazed at him startled. He had not been out of his bunk since his illness.

"What's the matter, brother?" I queried anxiously. "Why aren't you aft in your berth where you belong? We don't need help; we're getting along here beautifully."

"Where's the captain?" he asked, ignoring my questions. "I want him right away."

"Below there," I replied, pointing down the forepeak hatch. "He's inspecting the work on the bulkhead. Shall I call him for you, doc?"

Apparently too weak to speak a word more than he had to, Ambler only nodded. A little alarmed, I poked my head down the hatch into the dark peak tank and called out to De Long standing far below on the keelson. He looked up, I beckoned him, and he started cautiously to climb the icy ladder, shortly to be blinking incredulously through his frosty glasses at Ambler, even more astonished than I at seeing him out of bed. Ambler wasted no words in explanations regarding his presence.

"It's Danenhower, captain. I got up as soon as I could to examine him. His eye's so much worse to-day that if I don't operate, he'll lose it! So I came looking for you to get your permission first. You know how things stand with us all."

The captain knew, all right. It was easy to guess, looking into his harassed eyes as Ambler talked, what was going through De Long's mind—a sick surgeon, poor medical facilities, a leaking ship, and the possibility of having the patient unexpectedly thrust out on that terrible pack to face the rigours of the Arctic, where with even good eyes in imminent peril of freezing in their sockets at 50° below zero, what chance for an eyeball recently sliced open? All this and more besides was plainly enough reflected in the skipper's woebegone eyes and wrinkling brows. De Long thought it over slowly, then wearily shook his head.

"I can't give permission, doctor. It's not Dan's eye alone; it means his very life if we have to leave the ship soon. And since it's his life against his eye we're risking, he ought to have a voice in it. I can't say yes; I won't say no. Put it up to Dan; let him decide himself."

"Aye, aye, sir; I'll explain it to him." Dr. Ambler swung about, went feebly aft, leaving the captain and me soberly regarding each other.

"You're dead right, captain; nobody but Dan should decide. It's too much of a load for another man to have on his conscience if things go wrong."

De Long, abstractedly watching Ambler hobbling aft, hardly heard me. Without a word in reply, he turned to the ladder behind him, and with his tall frame sagging inside his parka as if the whole world bore on his bent shoulders, haltingly descended it. I looked after him pityingly. He had brought Dan, a husky, vital young man into the Arctic; now of all times, what a weight to have on his mind as Dan's life hung in the balance! Unconsciously I groaned as I turned back to thawing out my steam-line, and I am afraid that my mind wandered considerably for the next hour as I played a steam-hose back and forth along that frozen length of iron pipe.

I was still at it, and still not concentrating very well, when Tong Sing's slant eyes peered at me through the cloud of vapour enveloping my head and he pulled my arm to make sure he had my attention.

"Mister Danenhower likee maybe you see him, chief."

I shut off my steam-hose, nodded to the steward, started aft. If I could help to lighten poor Dan's burden any, I was glad to try. But what, I wondered, did he want of me—advice or information?

I entered Dan's room, sidling cautiously between the double set of blankets draping the door to shut out stray light. It was pitch-black inside.

"That you, chief?" came a strained voice through the darkness the minute my foot echoed on the state-room deck.

"Yes, Dan. What is it?"

"My eye's in horrible shape, the doctor tells me, chief. If it's anything like the way it hurts, I guess he understates it. What's happened to make it worse the last couple of days I don't know," he moaned, then added bitterly: "Most likely it's just worry. How do you think I feel lying here useless, not lending a hand, while the rest of you are killing yourselves trying to stop that leak and save the ship?"

I felt through the blackness for his bunk, then slid my fingers over the blankets till I found his hand.

"Don't let that get you, Dan," I begged, giving his huge paw a re-assuring squeeze. "We're making out fine with that leak. As a fact, we got it practically licked already. It wasn't much trouble."

"Quit trying to fool me, chief," pleaded Dan. "It's no use. Maybe I can't see, but I can hear! So I know what's going on around me. As long as I hear that hand-pump clanking, things are bad! And with the skipper's cabin right over my head and yours just across the ward-room and me lying here twenty-four hours a day with nothing to do but listen, don't you think I know when you turn in? And neither of you've turned in for a total of ten minutes in two nights now! Don't try to explain that away!"

I winced. Dan, in spite of the Stygian darkness in which he lived, had the facts. No use glossing matters over.

"Listen, Dan, I'm not fooling you," I answered, with all the earnestness I could muster. "It's true we haven't slept much, but we're both all right. And while things looked pretty bad at first, for a fact, we got that leak practically licked. Before the day's over, that hand-pump will shut down for good. Now forget us and the ship; let's get back to Danenhower. What can I do for you, brother?" I gave his palm a friendly caress.

I felt Dan's invisible hand twitch in mine, then close convulsively on my fingers.

"I'm in a tough spot, Melville. The doctor tells me if he doesn't operate, I'll go blind. And if he does, and I have to leave the ship before my eye's healed and he can strip the bandages, I'll probably die! And it's up to me to decide which. Simple, isn't it, chief?" Danenhower groaned. Had I not kept my lips tightly sealed, I should have groaned also at his pathetic question. With a lump in his throat, he added: "I don't want to go back blind to my f——," he choked the merest fraction of a second over the word, then substituting another, I think, hastily finished—"friends, but as much as anybody here I want to get back alive if I can. Honestly, chief, you won't fool a blind shipmate just to spare his feelings, will you?" He gripped my hand fiercely. "What're our chances with the ship? I've got to know!"

"The leak's licked, Dan," I assured him earnestly. "We won't sink because of that. But about what the ice is going to do to us, your guess is as good as mine. Seeing what she's fought off so far, I'd back the old *Jeannette's* ribs to hold out against the pack for a while yet."

"Thanks, chief, for your opinion." Dan pressed my hand once more, then slowly relaxed his grip. "I guess I'll have to think it over some more before I decide. You'd better go now; sorry to have dragged you so long from your work to worry you over my poor carcass."

I said nothing, I dared not, fearing that my voice would break. With big Dan stretched out blind and helpless on his bunk, invisible there, to me only a voice and a groping hand in the darkness, I slipped away silently, leaving him to grapple with the choice —to operate or not to operate—possible death in the first case, certain blindness in the second. And with the knowledge that however he chose, the final answer lay, not with him, but with the Arctic ice-pack. He must guess what it had in store for the

Jeannette with his sight or his life the forfeit if he guessed wrong. I went back to my own trifling problem, thawing out the steam-line.

Shortly afterward, Tong Sing came forward again, calling the captain this time, who immediately went aft. Whether Danenhower had decided or whether he was seeking further information, the steward did not know. I worked in suspense for the next hour till De Long returned. One look at his face informed me how Dan had decided.

"Well, brother, when's the operation?"

"It's over already, Melville! Successful too, the doctor says. I watched it and helped a bit. And, chief, I hardly know which to admire most—the skill and speed with which Ambler, weak as he was, worked, or the nerve and heroic endurance with which Dan stood it. He's back in his state-room now, all bandaged again. God grant the ship doesn't go out from under us before those bandages are ready to come off!"

Well, that was that. With a somewhat lighter heart, I resumed blowing steam on my frozen line. De Long crawled back into the forepeak to resume his study of the leak.

But my happier frame of mind did not last. If it was not one thing on the *Jeannette* to drive us to distraction, it was a couple of others. The captain soon squirmed back through the hatch with a long face to join me again beside the deck-pump.

"How much coal have we got in our bunkers, now, chief?" he asked.

"Eighty-three tons and a fraction," I answered promptly. I felt that I knew almost every lump of coal in our bunkers by name, so to speak.

"And what are we burning now?" he continued.

"A ton a day, captain, to run our pumps and for all other purposes, but as soon as that bulkhead's finished and the leak's stopped, we ought to get down to 300 pounds again, our old allowance."

De Long shook his head sadly.

"No, chief, we never will. The way the ship's built, I see now we'll never get that bulkhead really tight; she's going to keep on leaking and we're going to keep on pumping. But a ton of coal a day'll ruin us! By April, at that rate, the bunkers'll be bare. Can't you do something, anything, to cut down that coal consumption?"

I thought hastily. Our main boiler, designed, of course, for furnishing steam to propel the ship, was far bigger than necessary just to run a couple of pumps, and consequently it was wasteful of fuel. If pumping, instead of lasting only a few days more, was to be our steady occupation, I ought to get some set-up more nearly suited to the job. Before me in the deck-house was the little Baxter boiler I had rigged for an evaporator. That might run the forward pump. And looking speculatively aft through the deck-house door, my eye fell on our useless steam-cutter, half buried in a mound of snow and ice covering its cradle on the poop. There was a small boiler in that cutter. Perhaps I could remove it, rig it somehow to run a pump in the engine-room. And then I might let fires die out under the main boiler again and do the job with less coal.

Briefly I outlined my ideas to the captain, who, willing to clutch at any straw, gave blanket approval to my making anything on the ship over into what I would, so long as it promised to save some coal.

"Good, brother," I promised. "As soon as I get this pump running and knock off the hand-pump, I'll turn to with the black gang and try to rig up those small boilers so we can shut down that big coal hog. And even if we have to hook up Ah Sam's tea-kettle to help out on the steam, we'll get her shut down; you can lay to that!"

"I'm sure you will, chief," answered De Long gratefully. "Now is there any way we can help you out with the deck force?"

"Only by plugging away on those leaks, captain. We're making 3,300 gallons of salt water an hour in leakage; every gallon of that you plug off means so much more coal left in the bunkers."

"I well appreciate that, Melville. Nindemann and his mate are doing what they can with the bulkhead; I'm starting Cole and the deck watch to shoving down ashes and picked felt between the frames and the ceilings in the forepeak to stop the flow of water there. We'll get something on that leak, I don't know yet how much, but we'll never get her tight. I see that now."

And De Long, looking (though he tried to conceal it) as if that sight were breaking his heart, crawled back again to the freezing forepeak. I felt strongly tempted to seize him by the arm and start him instead for his bunk, but I was afraid he would urge the same on me and I had to get that line thawed and the Sewell pump going forward before I knocked off, so I let him go.

CHAPTER XXI

J ANUARY dragged away, followed in dreary succession by February, March, and April, and the *Jeannette* drifting aimlessly with the pack, was still solidly frozen in. Our lives were only a wearing repetition of what had gone before—fierce cold, alarms, the roaring and tumbling of the ice-pack, tremendous squeezes and pressures from the floes, and night and day the wheezing of the steam-pumps, pumping, for-ever pumping. It seemed almost a reasonable supposition to conclude that we must have the whole Arctic Ocean nearly pumped dry to judge by the length of time we had been at it and by the huge masses of ice banked up against our bulwarks and spreading out over the floes where the streams of sea water flowing from our scuppers had frozen.

A few minor triumphs and reliefs we had, but not many. In late January the sun came back over the horizon for the first time in seventy-one days, to reveal that we had all bleached strangely white in the long Arctic darkness. On the mechanical side, I had succeeded, after many heart-breaking disappointments, in supplant-ing the main boiler with the two little ones; and that, aided by the never-ending efforts of Nindemann in plugging leaks (which had cut the hourly flow nearly in half), had resulted in gradually re-ducing our coal consumption to only a quarter of a ton a day. We shot a few bears and a few seals, which gave a welcome variety to our diet of salt beef and tasteless canned meat; we even had hopes of knocking down some birds, but there we were disappointed.

"No, Melville," the captain gravely rebuked me, when empty-handed I returned to the ship after a February tramp over the floes and pushed my shot-gun disgustedly into the rack, "birds have more sense than men. No bird with a well-regulated mind would possibly trust himself out in this temperature."

On the debit side, the temperatures reached unbelievable depths. 57° below zero was recorded by our thermometers (the spirit ones, for the mercurial bulbs froze solidly at around −40°). The pack ice reached thicknesses of thirty-five and forty feet below the water where under-running floes, freezing together, consolidated into a kind of glacial layer cake. Contemplation of these formations, measurable whenever the floes near us cracked apart, gave a gloomy aspect to the ship's chances of ever getting free of the pack. And the irregular and formidable surface of the pack also gave us cause for thought, now that in the growing daylight we could see in what state the upheaval of January 19th had left the ice around us. Sledging across the pack was impossible; as soon might one think of get-itng from the Bronx to Brooklyn by dragging a team of dogs and a sledge over the Manhattan house-tops. Here and there, conditions were even worse. Sharvell, with the impressionability of youth, came in from an exploring trip with eyes popping to tell me:

"Say, chief, five miles north o' 'ere, the ice is standing in moun-tains 'igher nor our mast-'eads!"

"Yes, Sharvell, it's quite likely."

"Shall I tell the skipper, sir, or will you?" he asked anxiously.

"Why bother him about it?"

"If 'e knew, it'd save work, sir. 'E'd quit 'aving the bug-'unter clean an' mount that big walrus 'ead with the tusks that 'e's so busy fixing up. 'Cause when that ice gets to us, sir, we're through, an' it'll be a terrible lot o' work for us sailors dragging that 'eavy walrus 'ead over the pack. 'E better quit now, an' 'e will, sir, when I tells 'im abaht them mountains of ice!"

But I told Sharvell to forget it, for I doubted that with all his other worries, the captain would be much exercised over mountains five miles off.

Aside from the aspect of the ice, we had troubles closer home. Especially forward in the deck-house and crew spaces, the inside of the ship which now we had to keep above the freezing-point to save our pumps from damage, was damp and disagreeable beyond expression, with moisture condensing on all cold surfaces and drip-ping from the beams into the men's bunks.

Finally to deepen our gloom, Danenhower failed to respond

favourably to treatment, and the doctor had to perform several more operations on his eye, coming at last to the conclusion that Dan must, till we escaped from the ice, remain a chronic invalid confined in darkness to his cabin, with no great hope of saving his sight even should he then get back to happier surroundings and decent hospital facilities.

Oddly enough through all this, after the first week's struggle with the leak, we continued our scientific and meteorological observations. The captain clung to that routine as to a life-line, which perhaps to him mentally it was, constituting his solitary claim to conducting a scientific expedition. For of explorations and geographical discoveries there were none; on the contrary, instead of a steady drift northward which might uncover new lands or at least get us to higher latitudes, we shuffled aimlessly about with the pack, occasionally drifting northward for some weeks to De Long's obvious delight, only to have the drift then reversed and to his intense depression of spirits, to turn out some clear morning to find himself gazing once again across the pack at the familiar mountainous outline of the north side of distant Wrangel Land. But after March, even this sight of far-off land, depressing as it was from its associations, was denied us, for as the season advanced the pack, still zig-zagging over the polar sea as aimlessly as ever, failed to get quite so far south again; from that time on we saw land no more and the world for us became just one vast unbroken field of broken ice.

Only one hope kept us going. No one really knew what happened to that moving pack in summer-time—no one before us had ever wintered in it, involuntarily or otherwise. So we lived on in the expectation that as the days lengthened and the thermometer rose above zero, summer weather and the long days under the midnight sun would sufficiently melt the ice to break up the pack, and if by then we still had any coal left, permit us to do some little exploring northward before with bare bunkers we loosed our sails and in the early autumn laid our course homeward.

In that spirt, then, we cheerfully greeted the advent of May, and as if to justify our confidence, May Day burst upon us with gorgeous weather—no clouds, and glistening at us across the ice a brilliant sun which even at midnight still peeped pleasantly over the horizon, and a temperature which in mid-afternoon reached the unbelievable height of 30° F., only two degrees below freezing. We were positively hot. All hands (except, of course, Danenhower) turned out on the ice to bask in the sunshine, with the queer result that many of us came back aboard with our complexions sunburned to a fiery red and unable at first to believe it. Our hopes started to mount; if the sun could do that to such weather-beaten, frost-bitten hides as ours, what would it not do to the ice imprisoning us? Release was seemingly just around the corner of the calendar—by June 1st at the outside, say.

But meanwhile, awaiting that happy event, the captain prudently ordered (lest more casualties go to join the luckless Danenhower) that snow-goggles be worn on all occasions by all hands except when actually below on the ship.

So May moved along, made notable mainly by a positive flood of bears, which daily kept us on the jump. The bears, ravenous with hunger after a long winter, were attracted to the *Jeannette* by mingled scents, mainly canine, which to their untutored nostrils probably meant food. But we had long since lost any fear of ice bears, and the dogs apparently never had any, so the cry of—

"Bear ho!"

was the immediate signal for whoever had the captain's permission (which now meant practically anyone off watch) to seize a rifle from the rack placed conveniently at the gangway, and be off. We became so contemptuous of the bears that we chased them even with revolvers, and if necessity had arisen, would no doubt have done so bare-handed, for I have never seen a bear which would rush a man. Except when brought to by the dogs, with a man in sight all that ever interested the bear was to get behind the nearest hummock or into an open lead, where swimming with only his nose above water, he could escape the rain of bullets from our Remingtons and Winchesters. The vitality of the bears was amazing. Unless filled so full of lead that the mere weight of the bullets as ballast slowed them down enough for the dogs to bring them to a stand where a close-range shot into the brain finished them off, they usually got away.

We had queer experiences with the bears. On one occasion, exploring one of the narrow leads in the pack about a quarter of a mile from the ship, the captain was sculling unconcernedly along in the dinghy when he found himself facing an ice bear not a hundred feet off. Wholly unarmed, De Long regarded the bear with dismay. He could not run, for over broken ice he was no match in speed for Ursus; besides, he was in a boat, which prevented running away, for while the water was an obstacle to him, to the bear it was merely the most convenient means of transportation. Inquisitively the bear advanced; De Long, unable to do anything else, sat and stared, trying out the power of the human eye as a defence. The bear, only fifty feet off, still approached, sniffing curiously, and De Long, short-sighted though he was, said he could clearly make out where the short hairs ended at the edge of the bear's beautiful black nose. The captain quickly concluded there was nothing in hypnosis as applied to polar bears. So gripping his oar, prepared to fend off the bear should he approach closer to the boat, he sang out lustily:

"Ship there! A bear! A bear!"

At this the bear, more puzzled than ever, sat down on the ice to contemplate De Long and was still seriously thinking him over, trying to make him out, when a pack of dogs hove into sight from under the *Jeannette's* stern, followed by several seamen, and off lumbered the bear.

So long as we had the *Jeannette* under us, the plethora of bears meant at most only a break in the monotony of our existence and a welcome change in our salt beef diet. Should we have to abandon ship, however, they offered a ray of hope. For convinced now that we could never drag across the upheaved pack pemmican enough to keep us from starvation till we reached Siberia, we looked on the bears as a possible source of fresh meat on the hoof which we might with a little luck knock over as we went along and thus keep life in our bodies.

The only other spring-time event to compare with the bears was a brilliant idea which struck De Long.

While he never discussed his family with me or with anyone, De Long, alone among the ship's company which had sailed from San Francisco, had a wife and a child to occupy his thoughts. I have no doubt that frequently in the dreary months when I saw him, as I did one morning, abstractedly gazing out over the pack, his mind was far away from us, perhaps dwelling on that moment in the tossing whale-boat off the Golden Gate when Emma De Long had to the last possible instant clung round his neck in her farewell kiss. Drifting backward down the years from that, his thoughts on this morning evidently got to the days of his youth as an ensign aboard the U.S.S. *Canandaigua*. While cruising through the Channel ports, he had amongst the dykes and mills of northern France and Holland courted Emma Wotton, and as he thought of that landscape, so different from the ice-fields round the *Jeannette*, his keen mind saw a connection. He waved me to join him.

"Melville," he asked, obviously off again on the one ever-present topic, coal, "what'll you do to keep your pumps going when the coal's all gone?"

I pointed aloft.

"Cut down our masts and spars and burn them," I replied. "They're useless, anyway."

"And when they've gone too, what then?" De Long's clear blue eyes gazed at me fixedly, as if he had me there.

"Break up our bulwarks, the deck-houses, and the main deck, and shove those into the fires too. They'll all burn fine."

"And after that, what?" he asked relentlessly, puffing away on his ever-present pipe.

"I guess then we abandon what's left of the *Jeannette* and take to the ice, captain. I'll admit I can't keep any boilers going while I'm cutting the foundations out from under them for firewood. When the main deck's gone, I guess we're through."

De Long looked gravely at me through his glasses, bent his head a little to shield his pipe from the cold wind sweeping the deck, and irrelevantly asked me:

"Melville, have you ever been in Holland?"

"Why—yes," I mumbled, taken aback at his sudden change of front. "I guess it's tulip-time there now, captain. And quite a different scene from all this ice that's sprouting round us in the merry spring-time here. Why?"

"I was there in the spring-time once also," parried the captain. "Lovely scene. I just wonder if we couldn't make the scenery round here resemble Holland in the spring-time a little better. You remember the tulips, eh, chief? Do you by any chance remember anything else in the Dutch landscape—some windmills, for instance?"

And then a great light dawned on me. I looked at my captain with added respect. What did the Dutch have all those thousands of windmills for except to meet the same problem we faced—to pump water!

"Ah, you see it, do you?" asked the captain, gratified. "Melville, can you rig up a windmill here to run our pumps?"

"Can do, brother!" I exclaimed enthusiastically. "I'll turn to on it right away; before long you'll see a windmill going round here in the Arctic to beat the Dutch!"

This job was rather intricate for our facilities, windmills not being exactly in a sailor's line, but aided by Lee, machinist, and Dressler, blacksmith, we contrived it. Lee especially was a great help, which might seem somewhat surprising, for having been shot through both hips in the second day's fighting while helping Grant drive back Beauregard at Shiloh, Lee was rather slow and unsteady on his feet. But there was nothing the matter with his hands, and he soon had Dressler's crude forgings turned up in our lathe into a crank-shaft and connecting-rods, so that by the time Sweetman had made the wooden arms of the windmill, we were ready to go. Paradoxically, the one thing which on a ship we were best prepared to furnish, the sails themselves, failed to work well on our first trial. The mill occasionally hung on the centre because the heavy canvas sails sagged too much to hold the wind. Chipp, responsible for making the sails, watched them in pained silence, but having no canvas more suitable, soon rectified the matter in a novel manner. Sending Noros and Erichsen down on the ice, he had them collect some dozens of the empty meat-cans littering the ice-floes, and beating these out flat, he laced them together with wire, and soon had our mill-arms covered with fine metal sails! Impelled by these, our windmill, mounted on the starboard wing of the bridge, was soon rotating merrily and, connected by a special rig to a bilge-pump in the fire-room, was pushing overboard in grand style all our leakage. So well did it work that we quickly were enabled to shut down the steam-cutter's boiler, leaving only the little Baxter boiler going for distilling and in case the wind died down (which in the pack it rarely did) for unavoidable steam pumping.

So to our intense relief as spring drew on to its close, we got our coal consumption down again to 300 pounds a day, as it had been before that leak started to chew into our bunkers in such ravenous fashion. Which was a very fortunate thing for us, for with only sixty tons of coal left to go on, our days on the *Jeannette* would indeed otherwise have been numbered. Not least among the blessings which resulted was the improved cheerfulness of De Long at this success. He once more began to have some hope that when the ice broke up, we would have coal enough to do some exploring, so that he might again without too much shame on his return face our sponsor's sister, Miss Bennett, the ship's godmother, the "Jeannette" whose name we bore.

As the long days dragged out under the May sun, we eagerly watched the floes, noting with satisfaction the increasing number of rivulets coursing toward every crack and hole in the pack, and how under the intense sunlight, the cinders and ashes about the ship fairly seemed to burrow their way down into the snow. (Watching the striking manner in which everything dark soaked up the sunshine and settled, De Long half-humorously suggested that we all take a day off and pray for some miracle which might make all the snow and ice about us black and thus hasten its disappearance.)

And so we came to May 31, to our discouragement still held in the unbroken pack which, as measurements close about us showed, was still four feet thick. We decided to defer the day of our deliberation to July 1, giving the sun another month to work on the ice.

But to damp our spirits, June 1, the first day of summer as we reckoned it, opened in a snow-storm which continued through June 2 also, accompanied by a heavy gale which drove the snow, soft and mushy now, along in horizontal sheets.

When the snow finally ceased, the captain, optimistic again, began to prepare for the day of our release. First of all, fires were discontinued in the stoves fore and aft, thus saving a little coal. Next, all hands and the cook were turned to on knocking down our portable deck-house and clearing the main deck, so that looking like a ship once more, we might be able to spread sails and get under way when the wind served (provided, of course, the ice let go of us first). Several days' hard work accomplished this task, and with the topside shipshape again, we needed only to hang our rudder to be fully ready to go, but here again we had to wait on the ice which still clung solidly to our rudder post.

Below, I got my machinery and boilers in shape to move. With no fear of dangerous temperatures any more. I connected up all piping, moved the engines by hand, secured all cylinder heads, and filled both boilers to the steaming level (through the sea-cocks this time), and started generally to clean up the machinery spaces. For a small black gang, only six all told, this was slow work, so to avoid being caught with the pack suddenly parting and my machinery not ready to turn over, I pushed my gang hard. Consequently, I was doubly annoyed when I noted several times that Nelson Iversen, one of my coal-heavers and ordinarily a willing enough worker, showed decided signs of soldiering whenever my back was turned. I cautioned Bartlett who had charge of his watch, to get Iversen started, but after another hour, seeing he still tended to hide in the bunkers rather than scale rusty floor-plates, I yanked Iversen up sharply for it.

"Come to now, Nelse, and get behind that scaling hammer! Or will it take a little extra duty to keep you out of that bunker and on the job?"

Iversen, now that I got a closer look at him, looked queer in the eyes, so when, his slow mind having digested my statement, he finally answered, I was quite ready to believe him.

"Ay tank, chief, Ay work so hard Ay can. Ay ban sick man. My belly, she ache bad!"

"So, eh?" I said sympathetically. "Why didn't you tell Bartlett that an hour ago? Go up and see the doctor right away. What ails you, diarrhœa again?"

"No; de odder way."

"Constipation, huh? Well, you're lucky. On this bucket, that's a better thing to have than diarrhœa any day. Go up to the doctor and get some castor oil. And don't come back till it's quit working." I eased him over toward the fire-room ladder, and started him on his way toward Ambler.

But after a day had elapsed, I began to wonder whether the doctor's castor oil had somehow been affected by the cold or whether my coal-heaver had evaded swallowing his dose, for Iversen still showed the same tendency to shirk work and hide in the bunkers in spite of Bartlett's frequently breaking him out of there. So taking Iversen in hand myself, I escorted him up to the dispensary to see personally that there was no foolishness about his taking his medicine, and calling Tong Sing, I sent him off to find the doctor, who was out on the ice.

The minute Tong Sing disappeared, Iversen poked his head out the door, looked both ways quickly, then as if satisfied, hastily shut the door and to my complete bewilderment, stealthily approached me, cupped his hands over my ear and whispered:

"Chief, Ay no ban sick, Ay ban vatched! Dere ban mutiny on foot here!"

Mutiny? I stared at Iversen incredulously. The men were having a veritable hell in their life there in the Arctic, but what could they gain by mutiny? And who would lead it? For an instant I had a vague suspicion, but I resolutely put that out of my mind. Preposterous! I looked at Iversen intently. But there could be no doubt as to his sincerity. He was serious, all right.

I pushed him down into a chair, ordered sharply:

"Wait there, Nelse! I'll get the captain!" and closing the door behind me, I shot out of the dispensary and across the cabin to the captain's state-room forward in the poop. Fortunately, De Long was there, writing in his journal.

"Come with me, skipper. I want you to hear something. Right away!"

Puzzled unquestionably at my haste. De Long dropped his pen, put down his meerschaum pipe, stretched his six-foot frame up out of his chair, and reached for his parka.

"No, you don't need that, captain; just as you are. We're only going to the dispensary."

"Oh, all right. Who's hurt now?"

"Nobody, but come along!" I started back for the dispensary with De Long following, puffing leisurely at the retrieved meerschaum, which was his greatest comfort and his inseparable companion.

Iversen started up from his chair as we entered, saluted the captain, and again swiftly scanned the cabin outside before he closed the door.

"Now, Nelse, tell the captain," I said briefly.

Once more Iversen cupped his hands, whispered into the captain's ear. De Long's jaw dropped abruptly. His pipe fell from his mouth and only by a quick lunge did I save it from hitting the deck. But insensible to that, De Long, immovable, only started at Iversen, searching his face as I had done. Finally, he shook his head, muttered:

"It just can't be! Where'd you get this, Iversen?"

"Yah, cap'n. Ay tal you it ban yust lak Ay say! Ay ban asked to yoin. Ay no say, Yah; Ay no say, No; so Ay ban vatched clost. Dey kill me for'ard if Ay tal!"

De Long looked at me. I handed him back his pipe, which, wholly unconscious of his action, he took.

"What do you make of this, chief? It looks serious if Iversen's right!"

"Sounds crazy to me, but it might be so. Depends on who's in it and how many. The men are all armed, you know. The rifle rack's right at the gangway. Anybody can help himself, and lots of 'em are out on the ice, guns in hand this minute. But why

hey should want to mutiny, I can't see, unless the ice has affected
heir minds."

Shocked at Iversen's report; impressed by the gravity of the situa-
ion if Iversen were right, for there already with weapons in their
hands were the mutineers, the captain still looked sceptically at
my grimy coal-heaver. Why should his crew mutiny? But on the
other hand, what had Iversen to gain by lying about it? And
versen, a steady man, always carefully attentive to his duty, was
ust the type of seaman who might be trusted to stand with his
captain at all hazards.

"Well," said De Long grimly, "let's get into this! Now, Iversen,
who's behind it?"

But there the captain ran into a stone wall. Iversen, very nervous
now, became evasive, dodged the questions, and apparently in
mortal fear of his life, refused to name the mutineers, repeating only
over and over again how, for two days, he had been closely watched.
Threats, promises, got nothing more out of him. Finally, the cap-
ain, baffled, took a new tack.

"See here, Iversen, they can't hurt you, and nobody else'll get hurt
either if you tell. I can manage it then. There are eight officers
here; surely there are some of the crew will join us! I'll get all
the mutineers, if you'll name them, out on the ice on some pretext.
I don't care if they do go armed. Then we'll haul in the gangway
and from behind the bulwarks we can hold the ship! A couple
of nights freezing on that ice will bring them round, all right!
They'll come cringing back, hands in the air, begging to be taken
aboard. Out with it now! Who's the leader?"

Iversen, more nervous than ever, shuffled to the door, opened it
a crack to assure himself no one was eavesdropping outside, then
faced us, and tremblingly blurted out:

"Sharvell!"

An amazing change came over the captain. He dropped into a
chair, roared with laughter.

"Sharvell? That's rich! That lad? He's not even a man yet!
Nobody'd follow him in a mutiny any more than a child! Hah,
hah!" But abruptly he stopped laughing, for Iversen was now weep-

ing hysterically, tears running down his coal-stained cheeks. Soberly
De Long looked at him, then took me by the sleeve, pulled me aside
a little, and whispered.

"I guess the mutiny on the *Jeannette's* over, chief. I thought there
was somebody crazy in it, and now I know who. Send for the
doctor, quick! I'll stay here with Iversen." He started to light his
pipe again.

"I've already got the steward out looking for him, captain," I
replied. "Ambler ought to be here any minute. And I guess you're
dead right, brother! Poor Iversen!"

It was so. Immediately Surgeon Ambler came aboard we turned
the weeping coal-heaver over to him. An hour later, when, after a
careful examination, Iverson under Cole's surveillance had been
led forward, he confirmed our fears. Iversen, if not already insane,
was trembling on the border of it. Only observation over several
days could prove which. De Long, much relieved at first by
freedom from dread of any mutiny, was nevertheless badly enough
depressed by the doctor's report.

"First a blinded officer," he muttered, "now a crazy seaman!
What'll this ice do to us next?"

CHAPTER XXII

JUNE 21 came, the longest day in the year. Farther south, to ordinary people, that meant more daylight; to us, with daylight twenty-four hours every day, it meant only that the sun stood on the Tropic of Cancer, having reached his most northerly declination. Ruefully we considered that. The sun was as far north as possible, as high in our heavens as he would ever get, though even so, at noon he stood not so high, only about 40° above the horizon. We would never receive his rays any more direct; instead, from now on they would became even more slanting, and less hot as he went south. And we were still held in the ice. Our case for release began to look less hopeful, and we went around that day with cheerless faces. Long afterward, picked out of the Siberian snows, I salvaged the captain's journal and looking through it was particularly impressed by what he put down for June 21, 1880. So aptly did he express the situation and our feelings of desolation that day, that I repeat it here.

June 21, 1880. Monday.

"Discouraging, very. And yet my motto is 'Hope on, hope ever.' A very good one it is when one's surroundings are more natural than ours; but situated as we are it is better in the abstract than in realisation. There can be no greater wear and tear on a man's mind and patience than this life in the pack. The absolute monotony; the unchanging round of hours; the awakening to the same things and the same conditions that one saw just before losing one's self in sleep; the same faces; the same dogs; the same ice; the same conviction that to-morrow will be exactly the same as to-day, if not more disagreeable; the absolute impotence to do anything, to go anywhere, or to change one's situation an iota; the realisation that food is being consumed and fuel burned with no valuable result, beyond sustaining life; the

knowledge that nothing has been accomplished thus far to sav
this expedition from being denominated an utter failure; all thes
things crowd in with irresistible force on my reasoning powe
each night as I sit down to reflect on the events of the day, an
but for some still small voice within me that tells me this ca
hardly be the ending of all my labour and zeal, I should b
tempted to despair.

"All our books are read, our stories related; our games of ches
cards, and checkers long since discontinued. When we assembl
in the morning at breakfast, we make daily a fresh start. An
dreams, amusing or peculiar, are related and laughed ove
Theories as to whether we shall eventually drift north-east c
north-west are brought forward and discussed. Seals' livers as
change of diet are pronounced a success. The temperature of th
morning watch is inquired into, the direction and velocity of th
wind, and if it is snowing (as it generally is) we call it a 'fin
summer day.' After breakfast, we smoke. Chipp gets a soundin
and announces a drift east-south-east or south-east, as the case ma
be. We growl thereat. Dunbar and Alexey go off for seals wit
as many dogs as do not run away from them en route. Th
doctor examines Danenhower and Iversen, his two chroni
patients. Melville draws a little for this journal, sings a littl
and stirs everybody up to a realisation that it is daytime. Daner
hower (from his state-room) talks incessantly—on any and a
subjects, with or without an audience. The doctor moralise
between observations; I smoke; Mr. Newcomb makes his prepara
tions for dredging specimens; Mr. Collins has not appeared, h
usual hour being 12.30 in the afternoon. Meanwhile, the men hav
been set at work; a sled and dogs are dispatched for the day
snow for washing purposes. The day's rations are served out t
the cook, and then we commence to drift out on the ice to di
ditches, to look at the dogs, calculate the waste in the ice sinc
yesterday, and the probable amount by to-morrow. The dredg
is lowered and hauled. I get the sun at meridian, and we go t
dinner. After dinner, more smoke, more drawing, more singin
more talk, more ditch and canal-making, more hunting, more do

inspection, and some attempts at napping until four p.m., when we are all around for anything that may turn up. At 5.30 time and azimuth sight, post position in cabin, make chart, go to supper at six, and discuss our drift, and then smoke, talk and general kill-time occupations till ten p.m., when the day is ended. The noise subsides; those who can, go to bed; I write the log and my journal, make the observations for meteorology till midnight. Mr. Collins succeeds me four hours, Chipp him four hours, the doctor next four hours, Mr. Collins next six hours, I next two hours, Melville next two hours, and I end the day again, and so it goes.

"Our meals necessarily have a sameness. Canned meat, salt beef, salt pork, and bear meat have the same taste at one time as another. Each day has its bill of fare, but after varying it for a week we have, of course, to commence over again. Consequently, we have it by heart, and know what we are going to get before we sit down at table. Sometimes the steward startles us with a potato salad (potatoes now rotting too fast for our consumption), or a seal's liver, or a bear's tongue; but we generally are not disturbed in that way. Our bill of fare is ample and good, our water is absolutely pure, and our fresh bread is something marvellous. Though disappointed day after day we are cheerful and healthy, and—here we are."

And to all that I can fervently say "Amen!"

June on the whole was chilly and disagreeable. The temperature rarely got above 32° F., and yet in spite of that the ice did keep on wasting, from direct absorption of sunlight, of course. The ship came up somewhat through the softening ice to a lighter draft, owing to our considerable consumption of coal and stores since late November, when we were frozen in after our transit of the ice-canal. But as an offset to this cheering rise, she heeled gradually more to starboard, adding to our discomfort.

Meanwhile, De Long kept Dunbar, who naturally was a good walker, scouting far and wide over the pack looking for open leads, which might promise a break-up of the pack and a chance of escape through one of them. June 28, Dunbar, duck-hunting in the dinghy

N

in a little lead about a mile from the ship, came back in the late afternoon with thirteen ducks, but with what was far more exciting, the news that the lead had suddenly opened up, that he had followed it (open here and there to a width of half a mile) at least fifteen miles before turning round. And from there it still stretched northward as far as he could see!

De Long was immediately all excitement. If only we could get the *Jeannette* across that single mile of solid ice between, there was no telling how far north we might go along that lead! He dragged Chipp into his cabin and went over with him the possibilities of blasting out the intermediate ice. While Chipp was calculating how far our supply of gunpowder would take us, De Long, eager to size up the situation on the spot, hastily departed to examine the lead for himself.

About midnight he came back into the cabin, tossed his parka on to the table. Chipp, surrounded by a sea of papers containing his computations on the explosive powers of gunpowder, handed the captain a sheet containing his conclusions. De Long pushed it aside without even a glance.

"Never mind, Chipp, we won't need it. I got there just in time to watch that lead close up so tight you can't get a tooth-pick into it now! At least I had the melancholy satisfaction of realising that if the *Jeannette* had been there, she would in all probability have been in for a very fine squeezing!"

And so June ended. We were still in the ice. Danenhower, thin and bleached, was worse. Iversen seemed to be improved; while still occasionally hysterical, his delusions of mutiny were no longer obvious.

July came and went. We dressed ship on July 4 in a thick fog and a chilling mist. The flags came down at midnight (there was no sunset) all covered with frost. Rain, mist, and fog were general. Our hopes for what the summer sun would do for us began to fade. And even the few glimpses we got of the sun, instead of cheering the captain up, further irritated him. For De Long being now navigator and having finally after days of delay got a shot at the sun on meridian for latitude, hopeful that the drift had carried us north, glanced at his sextant only to exclaim in anguish:

"Look at that altitude! All the sun shows me is how much closer I'm getting to the South instead of to the North Pole! If ever a man had justification for profanity, this southerly drift is it! The Bible says that Job had many trials and tribulations which he bore with wonderful patience, but I'll bet *he* was never caught in pack ice! Nor drifted south when the wind was blowing north! But then Job's may have been an ante-glacial period!" De Long picked up his pipe and nearly bit the stem in half. But a puff or two of tobacco partly, at least, restored his equanimity. Putting his sextant back in its case, he remarked to Chipp, also engaged in shooting the sun:

"I suppose we might as well look at it philosophically. As Jack says, 'It's all in a cruise, boys; the more days, the more dollars!' "

July ended, and we were still in the ice. Such a miserable month we were glad to be rid of.

August opened. Looking back over two-thirds of the spent summer, with the highest temperature only 38° F. on the hottest day, all hands began to despair. So also, I think, did the captain, for he changed the schedule for taking meteorological observations, requiring them only once every three hours instead of hourly.

We came to the middle of the month, with the only change in our condition an increase in our heel to 7½°, a change indeed in something, but not an improvement. We began to get morose—summer was fast fading, we were not released, and our hopes of doing anything in 1880 or in any succeeding year were vanishing into space. I tried to cheer the mess up by singing (if I say it myself, for an engineer I have a very good voice), Irish songs and ditties having been my speciality since early in my Civil War days on blockade. Whether I cheered up anyone except myself with the sound of my voice, I do not know, but I did get some sullen looks for my efforts from Collins, who being Irish himself may have thought I failed to do justice to the songs of his native land. Collins (who also imagined *he* could sing) reciprocated by regaling us with melodies from *Pinafore*, then only two years old, but I thought he did the English far more violence than I did the Irish. In this conclusion, I have as independent evidence the reactions of New-comb, who, whenever I sang in the cabin, continued reading wholly oblivious of me, but whenever Collins opened up on *Pinafore,* immediately closed his book and remembered that he had a gull or a seal that required stuffing.

As August dragged along, the little pools of water covering the floes round about the ship now began to give us real cause for depression by freezing over at night with a skin of ice which failed to melt until the next noon. When that commenced, what chance was left for the sun to have any effect on floes still thirty and more feet in thickness?

And to add to our woes, we found that as a result of our southerly and easterly drifting, we had been steadily going backward. We were much closer to our starting point, Herald Island, in late August than we had been in early May. A whole summer's drifting in the pack, and for a Polar Expedition, we had got worse than nowhere!

Meanwhile, the wearing days crawled by and we chafed at our impotence—well, well-equipped and eager to do something, we lay idle. I could have chewed nails for a change; our captain was even more ambitious—entering his cabin one evening with a sketch for his journal he looked at me and asked abruptly:

"Know Hamlet, chief? No? Well, for something to do, like Hamlet I can say:

" 'Wouldst drink up eisel? Eat a crocodile? I'll do it!'

"And so I would, chief, if there were any eisel and a few crocodiles in our stores, and by so doing I could change our position to one of usefulness. Well, what have you got there for my journal? Another sketch of this eternal ice?"

August 31, the last day of summer, came and went. We were still fast in the pack. As a confirmation that summer was gone, we saw again that evening for the first time in months a faint aurora in the sky. De Long climbed to the crow's-nest with a telescope, took a look around. A desert of ice in all directions, nothing but ice, ice, ice! He came down from aloft, all hope of release gone. Calling the carpenter, he ordered him to commence preparing our portable deck-house for re-erection. Sending for me, he asked me to accompany him on a tour of the bunkers, to reassure himself, no doubt, on the coal question. Together, lighting our way with oil torches, we clambered through the dusty bunkers, the captain checking by eye my statements of the quantity in each one. Coming out, De Long musing over the figures, declared feelingly:

"God forbid anything happens to make us go back to steam-pumping. Only fifty-three tons of coal—an equal weight in diamonds would not tempt me to exchange! For that coal, chief, has got to last us through another winter in the pack!"

Only *another* winter? What reason, I wondered, had he for supposing the end of a second winter would find us any closer to release? But the pack had far from exhausted its versatility, as I soon enough found out.

CHAPTER XXIII

SEPTEMBER 1 came, and winter fell on us like a blanket. Snow, low temperatures, and the prompt freezing over of all stray pools with a coat of ice that failed to melt again gave the pack an immediate wintry appearance that only deepened as the month drew on. September 6, the anniversary of our being first frozen in, opened our second year in the pack, with the only change noticeable the fact that winter had set in earlier and harder. But of course our present position, a hundred and fifty miles north of that of the year before, might easily have accounted for that.

September drifted by. October came. The temperatures dropped into the sub-zero twenties. We noted only that we were less sensitive to cold than the year before—luckily for us, for apparently we were in for a worse freezing. All hands, officers and men, became more moody, less talkative. By now it was evident to even the dullest-witted that we might go on thus for ever in the ice-pack; that is, at least till death in one form or another—by starvation, when our food gave out; by freezing, when we exhausted our coal; or by the ice crushing our weak bodies at any time—put a period to our tale. To talk further about what the expedition would do when the ice released us seemed just a waste of breath. The ice was not going to release us.

Meanwhile, in spite of our dreary outlook, we had to stick to the ship, for what else could we do? But would the ship stick to us? What would the ice do to the *Jeannette* during this winter? Our memories of the horrors of the winter past were not reassuring.

The month drew along. We ate our tasteless food, we drank our distilled water, we kept ourselves alive. Two things only broke up our unvarying daily routine—Divine Service on Sunday, and the weekly issue (begun now for the first time on the cruise) on Wednesday of two ounces of rum per man. Jack Cole did not have to pipe long of a Wednesday afternoon to get the complete roster round the whisky barrel. But his long piping of a Sunday morning drew no such crowds. To Divine Service, conducted weekly in the cabin by the captain, came not a single seaman, and of the officers, just Chipp, Ambler, Dunbar and myself—a congregation of four only to hear George Washington De Long, acting-chaplain, feelingly invoke the blessing of the Almighty upon our enterprise and ask His mercy upon us—distressed, worn mortals, trapped in the Arctic wastes.

As October drew toward its close, distant rumblings in the pack, cracks in the floes roundabout caused by contracting ice, ridges of broken floes thrown up hither and yon, and the pistol-like snappings of shrinking bolts in our timbers, warned us of trouble. November came; we viewed its advent with trepidation, for the previous November had inaugurated our reign of terror. On November 6, the sun departed from us and the long Arctic night commenced, our second. It would be longer this time till the sun reappeared, ninety days or more instead of seventy-one, for we were farther north.

True to form, the thundering of the ice and the grinding of the pack recommenced as per schedule in November and the tremors coming through the thick floes shook the *Jeannette* as in a storm. But we were more calloused. Let the pack screech and roar! So long as nothing was happening close aboard we merely listened. Newcomb and Collins, however, who were more nervous than the rest, were for ever running up on deck at these shocks. They came back even more disturbed when they could see nothing than when moving ice within eyesight gave the explanation.

November drew along without visible disaster, but the dread and anticipation of terrors yet to come caused trouble in other ways. Newcomb, childish always, became mum as a clam at meals, and at other times talked to no one, except perhaps to Collins. Whatever De Long thought of this, he said nothing till one day passing through the taxidermy-room while Newcomb was mounting a crab, the latter stopped him, queried:

"Captain, will you ask Mr. Dunbar whether he saw that *Uria Grylle* he shot with his rifle yesterday, in flight?"

De Long, a little piqued perhaps at being thus asked by a very junior officer to serve as a messenger boy, said:

"Why don't you ask him yourself, Mr. Newcomb?"

"Because," replied our naturalist, "he has declined any relations with me."

De Long looked at him, puzzled.

"Declined? On a matter of duty? That sounds queer. I'll have to look into this." Poking his head into the cabin outside, he called the veteran Dunbar into the work-room, then closed the door.

"What's this, Dunbar, about your refusing to speak to Mr. Newcomb? He's just asked me to ask you a question about a bird you shot, because he says you won't speak to him."

"Let him ask," replied the ice-pilot. "I'll speak to him any time about anything in the line of duty. But not on other things; I despise that little Yankee pedlar and he knows it!"

"Come now, Mr. Dunbar," broke in the captain, "that's no way to talk about a shipmate. Don't lay too much stress on that little trading episode of Newcomb's with those Indians at St. Michael's; Mr. Newcomb did it only as a joke."

"A joke, eh?" burst out the angered whaler. "And I suppose it's a joke too, when he tries to write a letter home from Siberia, criticising his superiors, saying that you, the captain, are a profane Catholic and Melville's an atheist! A fine shipmate he is!"

De Long, at this unexpected personal turn, reddened, grew suddenly stern, gazed intently at Newcomb.

"What's this, Mr. Newcomb? I'm a Catholic, right enough, but I think no man can truly say I'm a profane one. Did you write such a letter, sir?"

"I did not!" said Newcomb promptly.

"I didn't say he *wrote* one," countered Dunbar. "I merely said he *tried* to. There wasn't any mail going, so I guess he didn't. But the little fool's too chummy with the men; it got out around the crew somehow that he was going to. That's where I heard it."

"Well, never mind about any scuttle-butt rumours, Mr. Dunbar. Mr. Newcomb says he didn't write such a letter, and that settles it. Now, Mr. Newcomb, I've noticed before your not talking to your fellow officers. Forget any such child's play, and you'll get along better."

"Don't I do my duty, sir?" asked Newcomb with apparent innocence.

"Yes, and I'll take good care that you continue to," responded the captain.

"Very well, sir," said Newcomb pertly. "If I do my duty, I must respectfully continue the privilege of maintaining this silence."

Nonplussed at this attitude, De Long looked at the infantile naturalist a moment, then gave up, turned on his heel and left. Needless to say, the ice-pilot promptly did likewise, leaving Newcomb in proud possession of his privilege of silence.

But this was only a beginning of increased ill-will in the mess, owing probably to the general state of ragged nerves. The very next morning, Dr. Ambler and Collins had a fierce set-to about the slamming of a door. It so happened that I was sitting in the ward-room, calmly reading a book, when along came Starr, the Jack-of-the-dust, to break stores out of the after-hold. He opened the ward-room door and fastened it back in order to roll a barrel through it, which he did. Just then four bells struck, and it being Ambler's turn to get the ten o'clock observations, the doctor drew on his parka and went out the opened door, followed soon by the huge Russian, who, sailor fashion, kicked the door to as he passed. The door closed with a bang, startling Collins, who as usual was asleep during the morning. Collins, grabbing a few clothes, shot by me out of his room, mad as a hornet. He never noticed Starr, who was still busy rolling his barrel forward, but spotting the doctor on his way up to the deck above, raced after him, seized him by the arm, and belligerently demanded:

"What d'ye mean by slamming the door like that? You know well enough I always sleep in the morning!"

Ambler looked at him in complete mystification.

"Why, what are you talking about, Mr. Collins? I haven't closed any doors, let alone slammed them."

"What d'ye think woke me up then? I'm not crazy! I heard you do it, and I'm damned sick of my being broken out of my sleep by you or by anybody!"

A dangerous hardness came into Dr. Ambler's usually soft Virginian voice.

"I tell you I didn't slam any door! Mr. Collins, do you mean to say to me that I lie?"

What might have happened, I don't know. I scrambled up the ladder, thrust myself between the two. From the glint in Ambler's eyes, it looked to me like bloodshed next.

"Hold on, gentlemen! Please!" I begged. "I know all about this. The Jack-of-the-dust slammed that door; Dr. Ambler didn't even notice it. Now look here, Collins! If you'd been faster on your feet, you'd have seen who did it yourself; and if you weren't so damned fast at taking offence at every little thing, you'd have rolled over and gone to sleep again without bothering about a little noise. Now apologise to the doctor and turn in again till you've slept off that grouch!"

Collins, very red in the face at having made a fool of himself, mumbled out some lame apology, which Ambler accepted without comment and departed to take the morning observations. I went below to continue my interrupted reading, but Collins, instead of turning in, moped about all day, no doubt trying to justify to himself his ridiculous conduct; very possibly wondering also whether I had not fabricated from the whole cloth that yarn about the Jack-in-the-dust to put him in a hole and figuring how he might reciprocate.

At any rate, at dinner that evening, he startled me by breaking his usual meal-time reticence and remarking as I was hacking away at the salt beef:

"There's old Melville, getting grey and bald over his confinement in the ice."

"No, Collins," I shot back, "my hair's no greyer than yours. And as for my baldness, I've suffered neither heat nor cold from it since I've been in the Arctic, but I will admit that if instead of being marooned here, we were off Saint Patrick's Land, where we could all be hunting now, probably I'd have a better time."

"So?" said Collins, instantly offended. "That settles it! When a man starts to get personal in his remarks, I don't have anything more to say."

"Personal? Who's getting personal?" I asked, perplexed, for if Collins's commenting on my baldness was not personal, what was? Then recalling my statement, I blushed myself, for in my haste in getting in my repartee I realised suddenly that my tongue had slipped. "Did I say *Saint* Patrick's Land, Collins? I'm sorry; I didn't mean to hurt your feelings. I meant *Prince* Patrick's Land, off to the north-east of here."

"Oh, no, Melville; I'm not a fool!" Collins blazed out, obviously certain now that I was altogether too facile in explaining away embarrassing situations. "When you said Saint Patrick's Land, you meant Saint Patrick's Land! And as for my grey hairs, I got them in honourable service you're completely ignorant of!"

Well, I thought to myself, where does he think I got mine? Surely the Civil War, which started me off on both my greyness and my baldness, was honourable service! But very prudently, I kept my thoughts to myself and my mouth shut. What was the use of further inflaming him? Quietly I bent my beard over my plate and resumed operations on my salt beef, while the rest of the mess, content to let the matter drop, wisely did the same and the meal closed in a tenser silence even than it had opened.

It began to seem now as if every little thing caused trouble. That night I had a remarkable dream, and there being so little to talk about that all hands had not heard discussed a hundred times over, I sprang it on the mess after breakfast, expecting to get a good laugh out of them.

"Say, mates," I began, "speaking of all the instruments we have to read on our meteorological observations, I had a grand dream about 'em last night. Want to hear it?"

"Guaranteed a brand-new dream, chief?" demanded Chipp. "If not, belay the story, for I dream about instruments every night now myself."

"Don't mind Chipp, chief. Shoot it!" encouraged the blindfolded

Danenhower from the foot of the table. "I can stand it, anyway; I don't have to read those instruments any more."

"Oh, it's new all right," I assured Chipp. "Stand by then. I dreamed last night I was old Professor Louis Agassiz himself, king of the scientific world, and without a stitch of clothes on, I was going down the middle of Pennsylvania Avenue on my way to the Smithsonian Institution, decked out with necklaces of hygrometers, bracelets made of thermometers, a belt like a South Sea hula-hula's grass skirt made up of mercurial barometers, and God knows what other instruments dangling from my fingers and my toes. And there I was dancing along through the heart of Washington with all those instruments on me clattering like castanets, offering to sell 'em to the crowd at only two cents apiece, but nobody would buy!"

Amid a gale of laughter from my mess-mates, I danced around my chair snapping my fingers, illustrating, then asked:

"Now, how's that for a dream, boys?"

"I think it's damned insulting to me and my profession, if you want my opinion!" broke in an unexpected voice.

Taken completely aback, I stopped dead in my dance and whirled about. There standing in his state-room door, watching me, was Collins, who, never on hand for breakfast, was at that time normally sound asleep. A dead silence fell on the laughing mess.

"And if you've got to try to make me look like a damned fool, Melville, with your jokes about nakedness and that my instruments are not worth two cents, wait till I'm off the ship!"

Slam! went the door, closing off our meteorologist, whom I had never even thought of in connection with my dream, from my sight. I sank back into my chair with a deep sigh. I couldn't even relate an innocent dream without offending the touchy Collins.

However, that was not the end of it, though I had hoped it was. The day itself wore along like all our other days, an utter blank, till about ten p.m., when with all hands about ready to turn in, the captain in his cabin sent for Collins, and as luck would have it, asked him, of all things, to bring a thermometer!

Collins went to fetch the thermometer, some special one, and

took it into the cabin. There was some conversation about thermometers which, the skipper's door being open, was faintly audible
to us in the ward-room, but to which I paid little attention, till,
the subject of thermometers evidently being now a raw one with
Collins, I heard him say in a loud voice:

"Captain, I wish the officers would treat me with the same courtesy
I try to treat them."

At that I pricked up my ears.

"What's the matter with you?" demanded De Long quickly.
"If you have any particular charge to make against any officer, make
it right now and I'll investigate it."

That was the last I heard, for the captain immediately closed his
door, wanting privacy of course for such a discussion.

"Well, here's where I have to explain even my dreams," I thought
to myself as I rolled into my bunk. "What a life!" Still, I managed to sleep that night with no more nightmares about thermometers to disturb me, and I woke in the morning quite refreshed.
Nothing happened during breakfast either, and I was beginning to
think that perhaps Collins was more of a man after all than the
night before I had given him credit for being, when a little after
eleven, while out on the ice for my regulation exercise, De Long
hailed me:

"Come here, Melville. I want to see you."

I went over. It was forty below zero, and, I thought, a devil of
a temperature in which to get hauled up over thermometers.

"Last night, chief," said the captain, starting mildly enough, "in
a conversation with Mr. Collins, he reported you to me for plaguing
him. I asked him what the trouble was, and he said that you were
always cracking jokes and singing Irish songs to make game of
him."

"What?" I mumbled half to myself, completely flabbergasted.
"Songs, in addition to thermometers?"

But the captain, oblivious of my interruption, finished decisively:
"Melville, you had better not sing any more."

"Why, captain!" I said in astonishment. "I don't think I should
be muzzled in this manner. There's no reason why I shouldn't sing

a song if I want to. It's my only relaxation. My songs don't disturb anybody."

"Collins says your Irish songs disturb him. Sing something else," ordered the captain flatly.

"But, captain, I can't. I don't know any other songs."

"Well, sing psalms then."

"Psalms? Me?" I protested. "Never! I didn't ship as a psalm singer!"

"Very well, chief; suit yourself," said the skipper with a note of finality in his voice. "It's a little cold out here to discuss the matter further. You had better stop singing altogether then," and leaving me badly upset at the idea of losing my one diversion, he walked off in the snow, resuming his exercise.

Naturally enough, I looked around the frosty field of ice to starboard of the *Jeannette*, which constituted our exercise grounds, for the cause of that muzzle the captain so unceremoniously had just slapped on me. A little way off was Collins, undoubtedly a witness to what had gone on, and in view of the extraordinary way sounds carried across the ice in that Arctic air, probably a willing enough auditor also. I strode over to him.

"Good morning, Collins."

"Good morning, Melville."

I was too hot in one way and too cold in another for any preliminaries. I jumped headfirst into my subject.

"The captain tells me you complained to him and claimed his protection against my jokes and my singing Irish songs and making game of you. Collins, that was neither upright nor manly!"

"Hold on!" said Collins. "I'll explain that thing."

"I don't want any explanations! It's plain enough what you've done. If you'd come to me like any shipmate should, and told me that my jokes and songs were disagreeable to you, I wouldn't have sung another song or cracked another joke. But your tale-bearing makes me sick! From now on, we're through! You keep to your side of the ship and I'll keep to mine! And don't you forget it!"

And from that day forward, I never spoke again to Collins nor he to me, except when I was told to carry him an order.

Our ward-room mess was now in a fine state of sociability. Danenhower, blinded, behind the bulkhead of his state-room talked almost incessantly to relieve his monotony, but nobody paid any attention to that as conversation. Dunbar wouldn't talk to Newcomb; Collins and I were not on speaking terms; Newcomb would not talk to anybody; Collins was nearly as bad, speaking pleasantly only to Danenhower; Chipp was naturally reticent and had little to say ever; Dunbar, much aged by illness, was taciturn as a result; the captain, weighted down by his responsibilities, felt compelled to maintain an extreme official reserve; and only Dr. Ambler and I were left ever to carry on a conversation like ordinary human beings. The ice was working on us, all right. A casual visitor, had one been able to poke his head through our door on the *Jeannette* at any meal, would have concluded that we were about to attend the funeral of some dear friend, and in that he would not have been far wrong; subconsciously we felt and acted as if we *were* going to a funeral, only it was—ours!

Matters in one direction at least soon came to a head with a rush. Collins, usually the last man out on the ice at eleven a.m. for exercise and the first man aboard at one p.m., when it ended, now began to comply with the exercise order even less cheerfully. As a regular thing he was considerably late in leaving the ship, and what was worse, he took an ungodly length of time, when he went aboard at noon to record the results of his midday observations in the log, to get back again on the ice with the rest of us. This quickly became such a flagrant flouting of the exercise order that while no one said anything about it, De Long could hardly overlook it and keep his authority with the rest of us. December 2 brought the end.

Collins, late on the ice that day as usual, went promptly aboard at eight bells to log the readings. When he failed to reappear after about three times the period required to note them down, De Long with an irritated look on his face boarded the ship. He glanced through the door, always open for ventilation purposes while the ship was cleared, into the cabin. There with his parka off, smoking his pipe, was Collins, leisurely writing in the log-book and carrying on an animated conversation through the opened doors and hatche with Danenhower in his state-room on the deck below. De Long further irritated at this confirmation of his suspicions, said nothin and returned to the ice, pacing nervously back and forth for anothe period long enough for anyone again to have logged the reading thrice over, and still no Collins! De Long reached the end of hi patience. With determination in his stride, up the gangway wer the captain and into the cabin. What happened afterwards, I g from Danenhower, who, an unwilling listener (unless he plugge his ears), was forced to take it all in.

Collins, at the stove, drawing on his gloves, was still talking wit Danenhower, when he looked up in surprise to find the skippe regarding him fixedly. De Long opened the ball.

"Well, Mr. Collins, has it required all this time for you to recor the 12 o'clock observations?"

Collins, a little nettled, replied:

"Well, sir, I hardly know the meaning of your question."

At this naïve disclaimer of comprehending simple English, D Long proceeded to explain in words of one syllable:

"The meaning of my question, sir, is this: Is it necessary for yo in order to record the 12 o'clock observations, to remove your coa light your pipe, engage in a conversation with Mr. Danenhowe and remain in the cabin for twenty minutes when you should l exercising?"

"Well," answered Collins curtly, "perhaps I might have done quicker, but I didn't know my minutes were being counted f me."

With difficulty swallowing the broad implication of spying co tained in Collins's last words, the captain said evenly:

"I have seen fit to issue an order that everybody should go on t ice from 11 to 1, and your coming in the cabin and remaining f twenty minutes is a violation I will neither submit to nor perm you to continue. I have noticed for several days that you we longer than necessary in logging the noon observations, and to-da I satisfied myself on the subject."

"Oh, very well," said Collins contemptuously, "if you are satisfie

f course I have nothing more to say. But you are doing me a
reat injustice!"

That was too much for De Long, who as captain prided himself
n even-handed justice for all hands. Whatever his ideas were
efore, he now changed his mind.

"Mr. Collins, as I have recently shown you, a representation to
ne of injustice has only to be made in proper language to secure
ou all the justice you want. But I do not like your manner or
earing in talking with me. You seem to assume that you are to
eceive no correction, direction, or dictation from me; that your
iew of an occurrence is always to be taken; and that if I differ
rom you, it is my misfortune, but of no importance to the result!"

At this Collins blazed up.

"Well, I don't like the manner you speak to me either, nor the
vay in which I am taken to task!"

De Long looked calmly at him.

"I am your commanding officer, Mr. Collins. I have a perfect
ight to say what I say to you."

But this Collins would evidently not admit. In a fiery tone, he
hot back:

"I acknowledge only the rights given you by Naval Regulations!"

That shot rocked De Long and he promptly flared up.

"Do you mean to imply that I am going contrary to Naval Regu-
ations?" he asked, outraged.

"I mean to say," said Collins flatly, "that you have no right to
alk to me as you do!"

De Long considered that carefully before speaking, then in as
fficial a voice as he could still muster he stated:

"I consider that by coming into the cabin as you did to-day, re-
noving your coat, lighting your pipe, and carrying on a conversa-
ion with Mr. Danenhower, you took advantage of the 12 o'clock
bservation to disregard my order in relation to the exercise."

"And when you say that," roared Collins, "I say it is *not* so!"

Amazed now by this open insubordination, De Long paused and
egarded the belligerent Collins with perplexity, puzzled by a situa-
ion so complicated that the like of it no commanding officer in naval

o

history had ever had to deal with. The captain finally decided t
try to calm Collins down, educating him a bit in naval manner
before finally admonishing him.

"Mr. Collins, great allowance has been made for your ignoranc
of Naval Regulations, your position in this ship, and your being s
situated the first time. But you must remember that the command
ing officer is to be spoken to in a respectful manner and with
respectful language, and you do not seem to attend to eithe
particular."

Collins rudely tossed this olive branch into the scuppers, so t
speak, by retorting truculently:

"I treat the commanding officer of this ship with all the respec
due to him as head of the expedition, but when he charges me with
violating an order, I say, I HAVE NOT!"

De Long accepted the challenge.

"Do you suppose you will be permitted to contradict me flatly in
that way, sir? Have you lost your senses?"

"No!" exclaimed Collins. "*I* haven't lost my senses. I know
what I say. And when you say I've violated an order, I say I hav
not!"

For the long-suffering De Long, that settled it. He rose,
dangerous coldness in his voice.

"Enough, Mr. Collins! You can't be properly dealt with in thi
ice. When we get back to the United States, I'll have you court
martialled! Meanwhile, turn in all your instruments, and perform
no further duty on this ship. You're under arrest!"

CHAPTER XXIV

DECEMBER was notable mainly for continued low temperatures, down around —50° F. We thankfully saw it slip away with nothing to remember it by save a minstrel show given by the crew to mark our second Christmas Eve in the ice. That this show was in any way memorable was mainly owing to my coal-heaver, comical little Sharvell, who rigged himself out as an attractive English miss in a sailor-made calico dress, a blond wig (originally the fibres of a manila hawser), white stockings, and low shoes. He provided so fair and alluring an imitation of something no sailor on the *Jeannette* had for a year and a half been within hail of, that the show was immediately a howling success, hardly needing the double ration of rum served out beforehand to make the audience not too critical of the performance.

Of Christmas Day itself, the less said the better. Our mince-pies were made of pemmican this time, the canned mince-meat having been all used up our first year's holiday. In spite of the brandy flavouring, there was probably not one of us who was not wondering with gloomy foreboding as he bit into his pemmican pie what, if anything, that crafty Chinaman, Ah Sam, would have left to sub-stitute for the mince-meat for our third Christmas in the pack.

And soon another New Year's Eve, with more minstrels; a little more rum; a fine speech to the crew by the captain, ending with the cheerily-expressed hope that before another New Year's Eve, we would all be back in our homes, saying to our friends with pardon-able pride:

"I, too, was a member of the *Jeannette* Expedition."

The only trouble with the speech was that no one, including the captain himself, in his heart really believed it. Then came midnight, with eight bells for the old year, eight bells more for the new one; and we soberly faced the year 1881. 1879 and 1880 had been heartbreakers for us. What had 1881 in store for the *Jeannette,* there in the Arctic?

To start with, it had January gales, bitter cold, and the usual thunderous uproar of the pack in motion, but fortunately never close enough to endanger us. And wonder of wonders, the discovery that the gales were mostly easterly, so that both by observation and by drift-lead, we found that at last we were going (when we went at all) steadily in one direction, north-west, and no longer endlessly zigzagging to and fro like a *Flying Dutchman* to the northward of Wrangel Land. And to lend a little further zest to this pleasing state of affairs, the month of January closed with the ship in latitude 74° 41′ N., longitude 173° 10′ E., a little farther west and three miles farther north than any of our previous peregrinations with the pack had ever got us. We began to take notice. Perhaps we were at last "going somewhere," although since the pack was moving with us, our scenery was changing not the slightest.

February arrived, and with it on February 5 came the SUN— a glorious sight to us after ninety-one long days of night! And never was he hailed by sun worshippers in ancient Persian temples with such sincere joy and enthusiasm as by the sadly-bleached and frozen array of care-worn, fur-clad seamen on the *Jeannette.* We streamed out on the ice with the temperature at 40° below zero to bask in the light, real daylight! of part of the sun's disk peeping over the horizon at us at noon. We thanked God for the sun's return, bringing to us once more the light to shine on our ship, still pumping away at our leak, but no more damaged than when he left us in early November.

The rest of February and all of March went by with no signs of let-up in our winter cold. A few more gales, seemingly worse than ever, buried the ship in fine snow, leaving only the three masts sticking up out of the white wastes to mark our position, but the wind continued easterly and we continued our north-west drift with our soundings steadily increasing also. Were we at last getting off the Siberian continental shelf into the deeper water of the open Arctic Sea? We hoped so, for deeper water meant greater opportunity for the ocean to break up the pack.

We drifted across the 75th parallel of latitude, for us a red-letter event. 75° North! It sounded much better. While not to be compared with the 83° North already attained along the Greenland coasts by other explorers, still it looked promising, and what a change from for ever shuttling back and forth between parellels 72° and 74° for twenty weary months!

In other ways, matters in March were worse. Several times, from the screeching of the pack, the cracking of the ice, and the severe shocks to the ship, we feared we were in for a repetition of some of our hair-raising experiences of the first winter, but each time the tumbling floes failed to come near us, and we thankfully heard the distant roarings subside. On top of that, the doctor, who all through the year had hopefully lanced and probed the abscess in Danenhower's left eye, found himself searching his soul as to whether he should undertake a major operation, but finally in view of all conditions, concluded he dared not. Dan, wan and emaciated from his long confinement, could at least see with one eye, his right, and that eye, while weak from sympathetic suffering with its mate, seemed now in less danger of becoming permanently affected. As a result, Dan with his left eye blindfolded and his right heavily shielded by a coloured lens, was occasionally allowed to walk over the ship, and even, when the weather was unusually favourable, permitted to grope his way round the ice alongside.

Aside from these matters, March brought us two other unusual episodes to break the monotony of our lives. The first was a bear, a she-bear as it turned out and recently a mother, which facts may have explained matters, for this bear, cornered by the dogs on top of a hummock near the ship, put up such a tremendous fight as we never saw before. The top of that hummock was a mass of flying fur and snarling dogs, the heavens resounding with howls, screeches, and roars, dogs leaping in to attack, only to be sent sailing right and left on their backs. Bear and dogs were out for a finish fight—savage teeth and lunging claws made a shambles of the ice on that hummock—how it might have ended was a question, for finally Nindemann, coming up, settled the battle with his rifle.

For us she turned out to be a very expensive bear; when we took

stock of casualties, we found one dog, Plug Ugly, dead; another
Prince, ripped open from back to shoulder; three more, Wolf, Tom
and Bingo, with gashed sides and stomachs; while Snoozer (the
captain's favourite) had his mouth considerably lengthened as by a
razor, and Smike was badly chewed in two places, not to mention
half a dozen other dogs licking minor injuries. Dr. Ambler put in
a busy day with thread and needle sewing up the wounded. When
it was all over, we had a badly battered pack of dogs who were quite
agreeable to crawling quietly off into the snow, by unanimous con-
sent suspending all hostilities among themselves.

The other deviation from monotony was the sudden interest
taken by our two Chinamen, Tong Sing and Ah Sam, in flying
kites. The steady and continuous east winds no doubt brought
to mind their opportunity, and soon Chinese kites in all shapes,
fashions, and colours—birds, flies, and dragons—were fluttering in
the breeze as tranquilly as if they were on the green banks of the
Yang-Tse-Kiang, instead of in the Arctic ice at 40° below. The
antics of our cook and steward with their playthings kept the crew
lining the *Jeannette's* rail watching them, in an uproar. But so
seriously did Sing and Sam take their pastime that when imperative
routine sent them back to cabin and galley, instead of winding in
their lines, they would tie their kite-strings to whatever was handiest
on the lee side, the shrouds, the davits, or the belaying-pins, till they
could emerge again and cast loose; and the captain believed that
had it been necessary, they would cheerfully have torn up their shirts
for kite-tails!

April 1 came, bringing with it by the calendar our spring and
summer routine, but no particular break in the weather, which on
April 16 was still —26° F., much worse than comparable temperatures of the year before; for us, certainly not a hopeful sign.

We soon had another bit of excitement; a few days later for a
while we feared that we had lost our entire commissary department.
Both Ah Sam and Tong Sing, armed with rifles, in the early afternoon went off hunting on the pack. When by seven at night they
had not returned, we became not only hungry but alarmed, and sent
out a searching-party. At nine o'clock they met the steward, Charley

Tong Sing, coming in alone, to tell a very involved story. A few miles from the ship, as he related it, he and the cook had picked up a bear-track, and with visions of more fresh bear to work on for dinner, they started eagerly in pursuit, after some miles coming in sight of the bear, which to their joy they found was being worried along by two of our best dogs, Wolf and Prince.

The dogs seeing reinforcements at last coming up in the form of our two Chinamen, and all hunters looking alike to them, promptly brought the bear to a stand by heading him off and snapping at his nose.

Running forward to get in a shot, Charley unfortunately slipped amongst some broken ice, and a piece of it fell on his back, holding him down, or he positively asserted, he would have killed the bear. Thus *hors de combat,* he lay while the dogs, no doubt thoroughly disgusted at such inexpert support, let the bear get under way again. By the time Ah Sam had managed to pry the ice off Charley and release him, neither bear nor dogs was in sight.

It being now at least six o'clock, both cook and steward came suddenly to the realisation that aboard ship, chow was way past due and held a council of war, the upshot of which was that the cook as senior officer present, ordered the steward to return to both cook and serve dinner while he, the cook, kept on to bag the bear. So there, safely back, was Charley Tong Sing, but where was our cook and where were our two best dogs?

De Long, having finally digested (instead of his dinner) this story in excited pidgin English of ice, bears, dogs, and Chinamen, looked at his executive officer in dismay. It was now dark, with considerable wind and drifting snow.

"Shall I send the searching-party out again, sir?" asked Chipp.

"What the use?" queried the harassed skipper. "A bear chased by dogs chased by a cook is too pressed for time to steer a proper compass course, so where should we look?"

We waited and worried till midnight, when that fear at least was allayed as Ah Sam, thoroughly exhausted, came stumbling up the gangway, and a more completely demoralised Chinaman you could never find. De Long personally made him drink half a tumblerful of whisky to bring him round, but he was completely incoherent and began to cry. When at last he was calmed a little, he related how he had continued to chase the bear, which the two dogs, to give him a chance, by fierce attacks managed occasionally to stop for a minute or two, but never for long enough for him to get within range. The dogs, Prince and Wolf, fighting desperately this way as the bear retreated, were both bleeding. Ah Sam says he followed the bear on a southerly course fifteen miles, determined to get him if he had to chase him all the way to China. Then by a particularly vicious onslaught, the dogs finally succeeded in holding the bear till Ah Sam could run up close enough for a fine shot. Raising his rifle, our cook took careful aim on the bear's head, and pressed the trigger, when horror of horrors, instead of hurting the bear, the rifle exploded in his hands! His morale completely shattered, poor Ah Sam sat down in the snow and wept, while the bear, still accompanied by Wolf and Prince, amazed no doubt by such weird hunting, but unwilling to give up, moved on over the pack and that was the last he saw of any of them. Still weeping, Ah Sam picked up the remains of his rifle and started home. How he ever found the ship again, he didn't know; it had taken him, walking continuously, until midnight. And there, indicating it with a hysterical wave of his hand, as proof of this wild story was the treacherous rifle!

We examined it curiously. Ah Sam had not exaggerated—the gun-barrel was torn to pieces; only a half length, cracked open, being left still attached to the stock. But to anyone used to fire-arms, the answer was simple. Ah Sam, in his long chase, must have let the muzzle slip into a snowdrift; the snow freezing solidly in the bore, had plugged it off, with the natural result that when he fired, there being no proper release for the exploding powder, it had promptly blown off the muzzle.

Dr. Ambler examined Ah Sam carefully for wounds; it seemed a miracle one of those flying rifle fragments had not cut his head off. But physically he had escaped unscathed; his demoralisation was wholly mental, owing to the way, in his efforts to provide roast bear for dinner, an unkind fate had treated him. Still weeping, poor Ah Sam was led off to his bunk.

CHAPTER XXV

APRIL drew toward its close, leaving us as a parting gift in latitude 76° 19′ N., longitude 164° 45′ E. Over 76° North, and with our drift increasing in speed weekly! We were on our way now with a vengeance, moving at last toward the Pole. A few more months like April, and we might find ourselves by the middle of summer across the 83rd parallel, to establish with the *Jeannette* at the very least a new record for Farthest North! The effect on George Washington De Long was magical—his shoulders straightened up as if he had shed a heavy weight, his blue eyes became positively cheery, new courage oozed from his every gesture —after twenty weary months of discouragement and defeat, our third year in the Arctic was going to redeem all and send us home unashamed!

May came. The temperature rose only a little, reaching zero, but we didn't mind that much, for in a few days we were nearing the 77th parallel. The captain's cheerfulness began to communicate itself to the crew and a livelier spirit became decidedly manifest in all hands, with one exception, that is. Collins, of course, was the exception. He, technically a prisoner awaiting court-martial, moped worse than ever; upset even more by the idea that now that he no longer had any active part, the expedition might really accomplish something. Physically Collins was not under restraint—no irons, no cell, not even restriction to his own state-room, let alone restriction to the confines of the ship. The captain had no wish to risk Collins's health by even such confinement as Danenhower was involuntarily subjected to. But relieved wholly of all duty and responsibility, Collins was in effect merely a passenger; his former work was divided between the captain, Ambler, Chipp and myself, throwing a heavier load on us, for the meteorological observations were religiously kept up. Indeed, with the ship at last rapidly changing position northward and westward, they were now increased. Still a member of the cabin mess, Collins ate with us, absolutely silent except for an ostentatiously polite "Good morning, captain," once a day, after which his fine oblivion respecting the existence of the rest of us was an excellent wet blanket on conviviality at meals.

But other things relieved the monotony of meals a bit. Ducks and geese began to show up overhead, flying some west, some north, and occasionally landing on the small pools near-by, formed by the continually changing cracks in the moving ice. Dunbar and Alexey knocked down some with their shot-guns. After our continuous diet of salt beef and insipid canned meat, rest assured we bit into those heaven-sent ducks avidly, though frequently sudden cries of pain as some *gourmet's* teeth came down hard on pellets of lead, showed that Ah Sam had been none too careful in extracting bird-shot before serving.

The weather warmed up a bit. The sun, though never high in the heavens, stayed above the horizon twenty-four hours a day, and even at midnight we began to see him, paradoxically enough, looking at us from due north, over the unknown Pole!

But as another paradox, now that winter was going and late spring and continuous daylight were with us, the doctor for the first time on our long cruise since the diarrhœa epidemic in 1879, began again to have a string of patients. Chipp, Tong Sing, Newcomb, Alexey, Kuehne, Nindemann, and, unfortunately, himself— all complained of general debility, cramps in varying degrees, and slight indications of palsy. Chipp, Tong Sing, and Newcomb, in the order named, were worst.

What was the trouble? The doctor, himself a minor sufferer, was able to work on his own symptoms as well as on those of the others in diagnosis. Naturally, since we had just come through the winter, scurvy was promptly suspected, but not a single evidence of the very obvious manifestations of that disease could the doctor find in anyone. This was some mental relief, for in the midst of all our other failures, De Long, Ambler, and I had taken considerable pride in having with my distilled water kept us free of that Arctic scourge and for a longer period than ever before in history.

But if it wasn't scurvy, what was it? Ambler racked his brains and his medical books, going over all possible diseases that cold, exposure, darkness, poor ventilation, depression, and our diet might have exposed us to, but to no result. The symptoms were none too obvious; he could lay his finger on nothing definite. Had we developed a new Arctic disease from our unprecedented stay in the ice? The surgeon could not say—only time would tell. Meanwhile, Chipp, the worst sufferer, decidedly thin and weak, was first relieved of part of his duties and then of all of them. The other victims were told to take things easier till they had recuperated.

But as the days dragged along, they didn't recuperate, they got worse. The doctor put Chipp on the sick-list and ordered him to bed; the same with Charley Tong Sing, whose case became even more serious. Meanwhile Ambler, suffering himself, was feverishly searching his *Materia Medica* for an antidote. But with no definite diagnosis of the disease possible, his search was fruitless. Ambler was nearly distracted, for no ailment arising from our manner of life fitted in with the vague symptoms. And then a severe attack of colic in Newcomb gave him a clue. He checked his medical books, checked the other patients, and with a grave face went to the captain to inform him that, implausible as it seemed, without question every man on the sick-list was suffering from acute lead poisoning!

That made the mystery even deeper. If lead poisoning, where was the lead coming from? Lead poisoning was normally a painter's disease, and not for month had any man on the ship touched a paint-pot or a brush. What then was the source? As the most probable cause, I had to direct suspicion at myself, for Bartlett, Lee, and I in making up our distiller piping joints, had for tightness wiped them all with red lead. Immediately, Surgeon Ambler, who had daily for a year and a half been testing the water for salt, tested it for lead. He found some insignificant traces, but it seemed hard to believe such minute quantities could cause us trouble. Still we had been imbibing that water constantly and the cumulative effect might have done it. While the problem of dismantling all the pipe joints and cleaning them of red lead was being cogitated, the captain went one step further—he ordered Ah Sam to discontinue for use in making tea and coffee, the pots which had soldered joints, and to replace them with iron vessels.

And so, all full of this lugubrious discovery as to what had laid up our shipmates, we met for dinner, a much reduced mess, with only De Long, Dunbar, Danenhower, Collins and myself present. Ah Sam, substituting for the deathly ill Tong Sing, served the meal —no bear, no seal, no ducks this time—just salt beef and the ever-present stewed tomatoes, our principal vegetable antidote for scurvy, the supply of which was holding out splendidly.

More quietly even than usual, dinner proceeded. I carved the salt beef, Dunbar ladled out the tomatoes. Ah Sam padded around the cabin with the dishes. Moodily we bent over our plates, and then an outburst, doubly noticeable in that silence, brought us erect.

"Bah!" burst out the semi-blinded Danenhower, spitting out a mouthful of food. "I don't mind breaking my teeth on duck, but who, for God's sake, shot these tomatoes?"

"Shot the tomatoes, Dan? What do you mean?" asked the puzzled skipper.

"Just what I say," mumbled Dan, trying more delicately with his napkin now to rid his mouth of the remainder. "They're full of bird-shot!"

I walked over and examined the tomatoes spattered on the table-cloth before Danenhower. Sure enough, there in the reddish mess were several black pellets of solder, looking remarkably like bird-shot! A light dawned on me.

"Ah Sam!" I ordered, "bring me right away half a dozen un-opened cans of tomatoes and a can-opener, savvy?"

"I savvy; light away I bling cans from galley," answered the cook, and in a few minutes dinner was suspended and forgotten, while the mess table was converted into a work-bench on which I opened cans and poured the contents into a large tureen. In every can we found drops of solder, mostly tiny! Evidently when the canned tomatoes were stewed before being served, the hot acid juices of the cooking tomatoes completely dissolved the fine lead pellets. They had never been noticed till a few drops large enough to escape com-plete solution had come through for Dan to bite on!

We called the sick doctor from his bunk. He promptly got his chemicals and then and there tested the hot stewed tomatoes already served for dinner. The percentage of lead in them was far above anything found in our water. No question about it now, the tomatoes were the cause—our mysterious lead poisoning was at last solved!

But the captain was still both perplexed and worried. Perplexed, because from the day we entered the ice, we had had canned tomatoes four times a week. Why hadn't we been poisoned before and why were some of us apparently still unaffected? He was worried, because if we gave up tomatoes, our last source of anything like vegetables, what (with our lime-juice now practically gone) over the long months to come was going to save us from scurvy?

Dr. Ambler quickly resolved both difficulties by pointing out that as for the perplexity, till May came, we had had tomatoes but four times a week, while since then we had had them daily, thus practically doubling the lead dosage and nearly as promptly starting the trouble. As for the reason why some were victims and some not—of the bad cases, Chipp, weak already from overwork and in poor condition, was a natural victim; Newcomb, little resistant to anything, another; as for himself and the bluejackets who were a little less affected, they were just somewhat more susceptible than some of the rest of us, but in a short time the lead would have got us all. Tong Sing's case, worse than anybody's, he had to confess he couldn't explain, but Ah Sam could and quickly did make it crystal clear:

"Cholly Tong Sing, he likee tomato! He eat plenty, allee same bleakfast, dinner, supper!"

All we need do to prevent scurvy was go back to the issue of tomatoes only four times a week, which quantity of lead absorption we had before apparently withstood. In addition we tried to reduce the lead still further by having Ah Sam carefully strain out and remove all pellets of solder *before* cooking, thus keeping the lead content down to the minimum, that is, whatever the cold tomatoes had already dissolved.

So with Ah Sam clearing away the mess of emptied cans, we went back to finish our dinner, lukewarm salt beef only now; silent again, wondering, if we had to stay in the Arctic another year, whether it was preferable to eschew the tomatoes and die of scurvy or to continue eating them and pass away of chronic lead poisoning.

The day dragged along. We were in the middle of May, it being the 16th. Our rapid drift continued through the afternoon, more westerly than northerly, but either was perfectly all right with us. The ice was "livelier," cracks and water leads showed up more frequently, the ship was often jolted by submerged masses of ice, and not so far away as earlier in the spring, high ridges of broken floes were piling up all around us. Then in the early evening after supper, from Mr. Dunbar, who more out of habit than hope, had crawled up to the crow's-nest for a look around, came the cry:

"LAND!"

And sure enough, there was land! Off to the westward lay an unknown island!

The crew of the *Jeannette* was delirious with excitement. Instead of ice, there was *land* to look at, something we had dully begun to assume had somehow ceased to exist on this globe. And we had discovered it! In exploration, our voyage was no longer a blank! In honour of that, Captain De Long immediately ordered served out to all hands a double ration of rum.

Not since March, 1880, when Wrangel Land last disappeared from sight, had we seen land. As yet we could not see much of this island, nor even make out its distance, but somewhere between thirty and seventy miles off it stood, in black and white against the sky and the ice, masked a little by fog over part of it. But our imaginations ran riot over *our* island! That must have been the land toward which the ducks and geese were flying, and when we got there, what a feast awaited us! Some eagle-eyed observers clearly spotted reindeer on its cliffs; others even more eagle-eyed plainly distinguished the bucks from the does! Our mouths, dry from chewing on salt beef, watered in eager anticipation.

De Long, positively glowing, hugged Dunbar for discovering our island, and looking happily off toward it, exclaimed:

"Fourteen months without anything but ice and sky makes this look like an oasis in the desert! Look at it, it's our all in all! How bears must swarm on *our* island, Dunbar! And if you want to tell me that it contains a gold-mine that'll make us all as rich as the treasury without its debts, I'll believe you! *Our* island must have everything!"

Even the sick, who came up on deck for a glimpse, were cheered by the sight, all, that is, save poor Danenhower, who nevertheless came up with the others, at least to look in that direction, knowing well enough that he alone of all of us would never see our island; that through the heavily smoked glass over his one remaining eye he could hardly see the bulwarks, let alone the distant island we had at last discovered!

Longer than anyone else, De Long stayed on deck that night, gazing off toward the island, criticising it, guessing its distance, wishing for a favourable gale to drive us towards it, and finally before going to bed, looking carefully again at it to make sure it had not melted away.

And when at last I dragged him below to rest, he murmured, knowing well the island could be only at most a little mass of volcanic rock:

"Melville, beside this stupendous island, the other events of the day sink into insignificance!"

For the next week, we drifted north-west with fair speed toward our island, with the water shoaling and the ice getting more active. By several bearings as we moved along, we discovered that when first sighted our island was thirty-four miles off. The question of making a landing began immediately to be debated, but obviously for the first few days, we were not yet at the closest point, so no decision was then arrived at. For the next three days, it blew hard, during which time we caught but few glimpses of our island as we drove north-west with the ice. When the gale abated on May 24, we got some sights and found to our pleased surprise that we were in latitude 77° 16′ N., longitude 159° 33′ E. 77° North! Another parallel of latitude left in our frozen wake; we were now moving steadily on toward the Pole!

But that was not all for May 24. Going aloft himself in the morning, De Long saw another island! Off to the westward it lay, closer to us even than our first island; and in addition, from all the lanes which had opened up in the pack, more water than he had seen since September, 1879. This second island, a little more calmly added to our discoveries than the first one, was a most welcome sight. The water, however, was nothing but a tantalising vision, for none of the lanes were connected nor did they lead anywhere, least of all toward our islands, both about thirty miles away from us and from each other.

Having two islands now on our hands, we could no longer refer to the first simply as *our* island, as we had before lovingly done in mentioning it, for was not the second equally ours? So it becoming necessary to distinguish between them in the future, De Long took thought like Adam of old, and named them—the first after our ship and our ship's godmother, Jeannette Island; and the second after our sponsor's mother, Henrietta Island. Having thus taken care of our sponsor's sister and his mother, De Long looked confidently forward to new discoveries on which he might bestow the name of our sponsor himself.

Meanwhile, the question of landing on either or both of our islands came again to the fore, the weather having cleared once more. Jeannette Island had dropped astern during our strong drift in the gale, while on Henrietta Island we were closing steadily. De Long decided therefore on May 30, six days after we had discovered it, to send a landing-party over the ice to take possession of Henrietta Island and to explore it.

The journey would evidently be a dangerous one over broken and moving ice, with, worst of all, the ship steadily moving with the ice away from the land. Most opinions were adverse to success, but Captain De Long ordered the trip, feeling that a knowledge of that island as a base to fall back on would be invaluable in case of disaster to the ship, and exceedingly desirous also of erecting a stone cairn there in which to leave a record of our wanderings and whereabouts (this, I think, though De Long never expressed it so, as a permanent clue to our fate should we be swallowed for ever by the pack threatening us).

Not as any compliment to me, but out of sheer necessity, De Long selected me to take charge of the expeditionary party and make the attempt to land. I was the only commissioned officer of the Navy available; Danenhower, Chipp, and Ambler were incapacitated in varying degrees; the captain himself, anxious as he was to have the honour of being first to plant our flag on newly discovered soil, dared not leave the ship to the only one other sea-going officer still on deck, the whaler Dunbar. So by a process of simple elimination, I was given the doubtful honour of leading. To help me were assigned Mr. Dunbar and four picked men from the crew—Quartermaster Nindemann and Erichsen, one of our biggest seamen, from the deck force; with Bartlett, fireman, and Sharvell, coal-heaver, from my black gang, the latter to act as cook.

With these men, one sledge, a dinghy to ferry us over any open water, provisions for seven days (including forty-two ounces of the inevitable lime-juice and eleven gallons of distilled water, but no tomatoes), navigating instruments, fifteen dogs, and the silken ensign which Emma De Long had made for the *Jeannette* as the particular banner to be used in taking possession, we shoved off from the vessel's side on May 31, cheered by all the remaining ship's company. Henrietta Island was twelve miles off over the pack, bearing south-west by west. The ship, to guide me in my return, hoisted a huge black flag, eleven feet square at the main.

Our sledge carried between boat and supplies, a load of 1,900 pounds, nearly a ton. With Dunbar running ahead as a leader to encourage the dogs and the other five of us heaving on the sledge to help along, it was as much as we could do to get it under way and moving slowly over the rough ice away from the ship. The harnessed dogs behaved as usual—they were not interested in any co-operation with us. In the first fifteen minutes, several broke out of harness and returned to the ship, there, of course, to be recaptured by our shipmates and dragged back to the sledge.

Of our terrible three-day journey over only twelve miles of live

P

ice toward Henrietta Island, I have little to say save that it was a nightmare. We made five miles the first day, during which we lost sight of the *Jeannette* and her black flag; and four miles the second. At that point, Mr. Dunbar, who had been doing most of the guiding while the rest of us pushed on the sledge to help the dogs, became in spite of his dark glasses totally snow-blind and could no longer see his way, even to stumble along over the ice in our wake. So we perched him inside the dinghy, thus increasing our load, and on the third day set out again in a snow-storm, guided now only by compass toward the invisible island. In the afternoon, the storm suddenly cleared, and there half a mile from us, majestic in its grandeur, stood the island! Precipitous black cliffs, lifting a sheer four hundred feet above the ice, towered over us; a little inland, four times that height, rose cloud-wreathed mountains, with glaciers startlingly white against the black peaks filling their every gorge.

As we stood there, awe-struck at the spectacle, viewing this unknown land on which man had never yet set foot, the silence of those desolate mountains, awful and depressing, gripped us, driving home the loneliness, the utter separation from the world of men of this Arctic island!

We were now only half a mile from the shore which marked our goal, but as we gazed across it, cold dread seized us. What a half-mile! The drifting pack, in which miles away the unresisting *Jeannette* was being carried along, was here in contact with immovable mountains which could and did resist. As a result, around the bases of those cliffs were piled up broken floes by the millions, the casualties in that incessant combat between pack and rock. While moving past between were vast masses of churning ice, for ever changing shape, tumbling and grinding away at each other as that stately procession of floebergs hurried along. And it was over this pandemonium that if ever we were to plant our flag on that island, we had to pass!

To get sledge, boat, and all our provisions across was utterly hopeless. So I made a *cache* on a large floe of our dinghy, stowing in it all except one day's provisions and most of our gear, raising an oar flying a small black flag vertically on the highest hummock of that floe as a marker. Next there was Dunbar. Terribly down in the mouth at having collapsed and become nothing but a hindrance, he begged to be left on the ice rather than encumber us further. But to leave an old man blinded and helpless on a drifting floe which we might never find again, was not to be thought of. In spite of his distressed pleadings, I put him on the sledge together with our scanty provisions and instruments, and then with a lashing to the neck of the lead dog who had no intention whatever of daring that devil's churn, we started, myself in the lead.

It was hell, over floes tossing one minute high in air, the next sinking under our feet. Splashing, rolling, tumbling, we scrambled from floe to floe, wet, frozen, terrified. Only by big Erichsen's truly herculean strength in bodily lifting out the sledge when it stuck fast did we get over safely. When at last, soaked and exhausted, we crawled up on the quiescent ice fringing the island, we were barely able to haul Dunbar, dripping like a seal, off the sledge and on to the more solid ice.

We paused there briefly while little Sharvell, his teeth still chattering from fright, clumsily prepared our cold supper. Then marching over the fixed ice, I as commander first set foot on the island and in a loud voice claimed it as a possession of the United States. I invited my shipmates ashore, and in a formal procession led by Hans Erichsen (who as a special reward was carrying our silken ensign) they landed also on the island, where Erichsen proudly jammed the flagstaff into the earth.

With a few precious drops (and precious few) of medicinal whisky, I christened the spot HENRIETTA ISLAND, after which we six sick seamen drank the remainder of the medicine in honour of the event, and then revelling in a brief tramp over real earth for the first time in over twenty-one of the longest months men have ever spent, we hauled our sleeping-bags about our weary bones and lay down, at last to rest again on *terra firma*.

At 10 a.m. we woke, startled to have slept so long, for we were not to stay on the island longer than twenty-four hours. On a bold headland near-by, we built our cairn, burying in it two cases, one zinc and one copper, containing the records with which Captain De

Long had provided us. This promontory, Mr. Dunbar named "Melville Head" in my honour, but after considering its bareness of vegetation, I decided "Bald Head" was more appropriate, and so entered it on the chart I now proceeded to make.

With Bartlett and Erichsen reading instruments while I sketched, we ran a compass survey which took all day. From the high headlands, the *Jeannette* was plainly visible in the ice to the north-east, a black speck against the white pack, but we paid little attention to her, being anxious only to finish. While this was going on, Sharvell and Nindemann searched the valleys, shooting a few of the birds nesting in great profusion among the rocks. But aside from the birds they saw no other game—no bears, no reindeer, no seals—not a trace of animal life on that island.

In the early evening, our survey finished, we harnessed again our staked-out dogs, furled our banner, and started back.

Our retreat through the roaring ice about the island we found even more difficult than our landing. On one small floe, rounded like a whale-back, we took refuge for a moment in that cascading ice. We clung on in terror when it began rolling beneath us, evidently about to capsize. That to our dismay it finally did, but providentially we were scraped off as it went over on to the main floe. From this more solid footing we dragged up out of the icy water by their harnesses the drenched dogs and the even more drenched Dunbar clinging to the submerged sledge.

Back once more on ice moving only as part of the great Arctic pack, we breathed a little more freely, shook ourselves like the dogs to get rid of surplus water before it froze on us, and headed for the spot toward which I figured our abandoned boat had drifted. There was nothing we could recognise, there were none of our previous tracks we could follow; the arrangement of that pack had changed as completely as a shuffled deck of cards. Amongst high hummocks we could see but a little distance and I was becoming thoroughly alarmed at the prospect of never finding our boat again. Then with the weather clearing a bit, from the top of the highest hummock around, Erichsen spied in the distance the oar marking our boat. We hastened toward it, truly thankful, for we had already made away with the single day's rations which we had carried with us, and had no longer a bite left to eat.

For two days in miserable weather we stumbled back toward the ship, steering a compass course through continuous snow. To add to our troubles, Nindemann came down with severe cramps (lead, of course) and Erichsen, who since Dunbar's collapse had been guiding the dogs, with snow-blindness. So pitching our tent in the snow, we camped our second night, while I dragged out the medicine-chest with which I had been provided by Dr. Ambler and began to read the directions. The remedy for cramps was "Tincture of capsicum in cognac." Henrietta Island having seen the last of the cognac, the best liquid substitute available in the chest appeared to be a bottle of sweet oil, which I drew out, together with the bottle marked "Tinc. capsicum."

My own fingers were cold and numbed, so Erichsen who wanted some of the sweet oil to rub on his chafed body which he had stripped for that purpose, volunteered to draw the corks for me. First pouring some of the sweet oil over his hands to soften them, he pulled the second cork, but so clumsily with his frozen paws that he spilled a liberal portion of the tincture of capsicum over his badly chapped hands to discover promptly that compared to tincture of capsicum, liquid fire was a cooling, soothing lotion!

Startled, Erichsen involuntarily rubbed the mixture on his bared rump and immediately went wild. To the intense interest of his shipmates, down went Erichsen into the snow, trying to extinguish the burn, wiggling his huge form like a snake on fire. Little Sharvell, solicitously taking his arm, piped up:

" 'Ere, matey, let me lead you to a 'igher 'ummock! Bli' me if I don't think ye'll soon melt yer way clean through this floe!"

Nindemann began to laugh so hard at this that he completely forgot his cramps, while Dunbar, between his own groans, sang out cheerily:

"Hans, are ye hot enough yet to make the snow hiss? If ye are, when we get back, the chief can put out the forecastle stove and use ye for a heater!"

Amid the general merriment, joined in by all hands except poor

Hans, big Erichsen finally managed to cool himself down in the snow enough so that he could stand an administration of pure sweet oil to the affected parts. Carefully applied by me, this soothed him enough to permit his dressing again, and with most of us in a hilarious frame of mind, we slid into our sleeping-bags.

Next day, our sixth since departure, we set out again at 3 a.m., and mirth having proved a better cure than medicine, with all hands in fair shape except Dunbar, who still had to ride the sledge. Within an hour we sighted the ship. This cheered us further. And the dogs recognising the masts and realising that at last they were pulling toward home, for the first time put their hearts and shoulders into the job. Over bad ice, we made such excellent progress that by 6 a.m. we were within a mile of the ship, apparently without having been sighted from there.

At this point, I ran into an open water lead with running ice, and unable to find a detour, had determined to launch the dinghy and ferry across when a sledge runner gave way and left us flat in the snow. We repaired the runner, but it was evident that it would never carry all the weight again. So I unloaded the boat, ferried the sledge across, and then sent it ahead with Dunbar only on it while Sharvell and I stayed behind with the dinghy and all the rest of our sledge-load of equipment.

We were all soon sighted and a party came out from the ship. There on the ice, Dr. Ambler met me, and undemonstrative though he was, so overjoyed was he at our safe return that he gave me a regular bear hug.

Approaching the gangway, we caught sight of Captain De Long, enthusiastically waving to us from the deck, running up the ladder to the bridge for a better view. Then to our horror we saw him, absorbed only by our progress, step directly into the path of the flying windmill! In an instant, before anyone could cry out in warning, down came one of the huge arms, whirling before a fresh breeze, hitting him a terrific blow on the head and sending him reeling backwards down the ladder!

Fortunately the quartermaster caught him, breaking his fall, but Ambler and I, forgetting all else, rushed for the gangway, arriving on deck to find the captain crawling on hands and knees, stunned and bleeding from a great gash in his head. Ambler hurriedly bent over him, carefully feeling his skull, and announced thankfully there was no evident fracture. He helped the semi-conscious captain to his cabin, where he immediately went to work stitching up a deep four-inch long wound. By the time this was done and the bandages applied, De Long at last came out of his daze. But calloused as I was by war and many hardships, it nevertheless brought tears to my eyes when his first question after his fluttering eyelids opened on the doctor bending over him, was not about himself but a faint query:

"How about Melville and his men, doctor? Are they all safe?"

CHAPTER XXVI

O N June 5, 1881, a Sunday morning, we got back to the *Jeannette*. In the early afternoon, in honour of our safe return, De Long with his eyes hardly visible through his bandages, conducted a Thanksgiving Service, attended only by Ambler and myself, for the other two usual members of the congregation, Chipp and Dunbar, were both on the sick-list. In further celebration of the event, the captain ordered in the evening the issue of a double ration of whisky forward, which ceremony conducted in the forecastle by Jack Cole drew a somewhat larger attendance, I believe.

Our sick-list was now considerable—Danenhower, Chipp, Newcomb, Dunbar, and Alexey, with the skipper himself really belonging there, but nevertheless permitted by the doctor to be up so long as he stayed off the ice for a few days till his cut had a fair chance to start to heal. Chipp, Newcomb, and Alexey were still badly off from lead poisoning, but Tong Sing, our steward, had recovered sufficiently to go back on duty and was now mainly engaged in tending the sick when not actually serving.

From this unsatisfactory state of our personnel, I turned my attention after a week's absence once more to the *Jeannette* and what was going on round her. Henrietta Island was rapidly dropping abaft our beam as we drifted westward past its northern side and it was evident that we would soon drop it out of sight. Jeannette Island had already vanished from our world.

But the action of the ice about us attracted most attention. Not since November, 1879, had we seen so much moving ice near the ship, the effect undoubtedly of near-by Henrietta Island. The day after my return, we found our floe reduced to an ice island about a mile one way and half of that the other, with ourselves about a hundred yards from the western edge, while all about us was a tumbling procession of floebergs, shrieking and howling as they rolled past. Leads opened and closed endlessly in the near distance with ridges of broken floes shooting thirty feet above the pack. The roaring of the breaking floes sounded like continuous thunder. And in all this turmoil our ice island with the *Jeannette* in it moved majestically along. Meanwhile we from our decks regarded it, thankful that our floe was not breaking up to crush our ship and leave our heavy boats and sledges to the mercies of that chaos, a half-mile of which with a sledge lightly loaded only, off Henrietta Island we had barely managed to survive.

Another day passed, leaving the island in our wake. The moving ice closed up again with long rows of piled-up floes all about us, one huge ridge of blocks seven to eight feet thick riding the pack not a hundred and fifty yards away from our bulwarks. And yet one more day and the captain got a sight, showing we were going due west at a fair rate, which if continued, unless we turned north, would ultimately bring us out into the Atlantic, though the chances seemed better for a resumption of our north-west drift toward the Pole. But toward either of these, now that we had some discoveries to add to the world's charts, we looked forward hopefully. At any rate, since we had to leave the matter to the pack, for the present our motto was obviously "Westward ho!"

June 10 came with our drift still steadily westward, clear weather, and the temperature about 25° F., well below freezing though above zero, which for us made it very pleasant weather. Alexey came off the sick-list, and so also did Dunbar; leaving only Chipp as a bedridden case, and Newcomb, up but acting as if he were exceedingly miserable, which I guess he was. Danenhower, permanently on the sick-list, was allowed on deck an hour a day for exercise that the doctor hoped would gradually restore his health and save his one good eye, which now showed some signs of getting over its sympathetic inflammation. During these hourly periods, Dan was sternly ordered to keep in the shade and wear his almost opaque shield, but unfortunately our over-bold navigator stepped out into the sun and pulled aside the glass, attempting to get at least one decent look around. Instead he had an instant relapse of his inflamed eye which nearly drove both Ambler and the captain wild.

Fortunately, the captain had had all his bandages save one small one removed from his injured skull by now, or I think he would have ripped them off in his attempts to tear his hair over the results of Dan's reckless disobedience.

Except for this unfortunate mishap, June 10 passed away pleasantly enough. With no more thought than that it was just another day in the pack, most of us turned in at 10 p.m., concerned only about whether our drift next day would continue west or change to north-west. But I, having the watch from 9 p.m. to midnight, remained on deck. At 11 p.m., I was disturbed by a succession of heavy shocks to the ship which increased in frequency till as midnight approached there was such a thumping and thundering of cracking ice about us and so much reverberation as the shocks drummed against our hollow hull, that the uneasy deck beneath me quivered as I had not felt it since two years before when we had been under way with all sail set. So violent was this disturbance that De Long, asleep below, lost all thought of rest, pulled on his clothes and scrambled on deck to see what was up.

With the sun even at midnight above the horizon, he had little difficulty seeing, and, of course, none at all in hearing. About eighty yards from us, a lane had opened in the pack some ten feet wide, while all about us as we watched, cracks were zig-zagging across the surface of our floe to the accompaniment of thunderous detonations as the thick ice split. And all the while, the heavily listed *Jeannette,* still fast in the ice, rocked in her bed as in an earthquake.

For ten trying minutes this went on, and then with a terrific report like a bomb exploding, the floe split wide apart beneath us, the *Jeannette* lurched wildly to port and suddenly slid out of her cradle into open water! There she rolled drunkenly for a moment, till coming finally erect she lay free of the ice at last in a swiftly widening bay!

So rapidly did all this happen that the skipper, clinging to the rail of his reeling bridge, saw the situation change from that of a ship frozen into one under way before he could give a single order.

But immediately after, with the ship still rocking heavily:

"All hands!" was echoing fore and aft, and I rushed below to close the gates in our watertight bulkheads and stand by my steam-pumps, not knowing what effect this sudden release of our bow from the ice might have on that leak we had been pumping, so it seemed, for ever. Paradoxically, however, the leak immediately decreased, probably because our freed stem floated several feet higher than before, so I returned quickly on deck to find the crew under the captain's directions busily engaged in preparing to re-ship our long-disused rudder. This, delayed by frozen gudgeons, took some hours. But when it was completed, and everything meanwhile had been cleared away from booms and yards for making sail, the *Jeannette* for the first time since 1879 (though we never saw the irony of that till later) was again ready to manœuvre as a ship.

Amidst the hoarse orders of the bosun and the noise of seamen clearing running rigging and scrambling out on frosty yards to loose the preventer lashings on the long-unused sails, I climbed to the bridge. There I found De Long calmly smoking his pipe while he eyed the smooth black water in our bay, now perhaps a quarter of a mile wide between the separated edges of our late island.

"Shall I fire up the main boilers, captain, and couple up the propeller shaft?" I asked anxiously.

"How much coal have you left, chief?" he countered.

"Only fifteen tons, sir."

Fifteen tons. That would keep us going only three days normal steaming. De Long thought a moment.

"No, chief, don't light off. There's no place for the engines to take us, anyway, and we might burn up all our fuel just lying here. Save the coal; we may need it to keep us from freezing next winter. We'll make sail if we have to move, but just now, all we can do is get some lines ashore and tie up to that starboard floe, till we see what the pack is about."

So instead of trying to move, Cole ran out the lines to ice-anchors on our bow and quarter and we moored to the floe.

Then began a desperate fight with De Long struggling to save his ship should the ice close in again before it broke up completely and let us escape. A measurement near-by showed the ice sixteen feet thick; deeper than our keel. If the pack, pressing in on us now, got a fair grip on our sides, we should be squeezed between thicker ice than ever before we had been, in a giant nut-cracker indeed. But what could we do about it? The water lead was short, there was no escape from it ahead or astern. Just one chance offered itself. A little ahead of where we lay, on our port bow was a narrow canal joining two wider bays in the parted pack. If we could only fill that canal up with heavy floes, they might take the major thrust of the closing pack, thus saving us from the full pressure. Savagely the men on watch turned to and fought with lines and grapnels, hooking loose floebergs everywhere and dragging them through the water into that canal, anchoring them there as best they could.

We had made fair progress on filling the gap, when at 7.30 a.m. the ice started to advance. The sight of that massive pack slowly closing on us like the jaws of doom, quickened our muscles, and we strained like madmen shoving drifting ice into the opening ahead. Just then, as if playing with us, the pack halted dead, giving us a better chance to finish the job.

De Long came down off the bridge to encourage the men with the grapnels. Standing on the edge of the canal, directing the work, was our ice-pilot. Approaching him along the brink of the pack, the captain looked down through the cold sea at the submerged edge of the floe, the blue-white ice there glimmering faintly through the water till lost in the depths; then he looked back at the *Jeannette* with her tall masts and spreading yards erect and square at last across the Arctic sky, while her stout hull, stark black against the ice, seemed grimly to await the onslaught.

"Well, Dunbar," asked the skipper, "what do you think of it?"

Dunbar, worn and dour, had his mind made up.

"No use doing this, cap'n," he replied dully, indicating the men heaving on the grapnel lines. "Before to-night, she'll either be under this floe or on top of it! Better start those men, instead o' hauling ice, at getting overboard the emergency provisions!"

De Long shook his head. He couldn't agree. In terrible winter weather, the sturdy *Jeannette* had often beaten the pack before; he couldn't believe that she would fail us now.

At ten o'clock, the ice started to advance once more. Our job in plugging the canal was finished. We had done all that man could do. Now it was up to the *Jeannette*. But as we watched that pack come on, flat floes and tilted floebergs thick and jagged, urged forward by endless miles of surging ice behind, our hearts sank. In spite of our thick sides and heavy trusses, the contest between hollow ship and solid pack looked so unequal.

On came the pack. The bay narrowed, thinned down to a ribbon of water on our port side, vanished altogether. The attacking floes reached our sides, started to squeeze. The *Jeannette,* tightly gripped, began to screech and groan from end to end. With bow lifted and stern depressed, she heeled sharply 16° to starboard, thrown hard against the floe there, while we grabbed frantically at whatever was at hand to avoid being hurled into the scuppers. Then to our intense relief, the ice we had pushed into the canal ahead came into play, took the further thrust, and stopped the advance, so that for the moment everything quieted down, leaving our ship in a precarious position, but at least intact. Our spirits rose. Perhaps we had saved her!

Thus we lay for two hours till eight bells struck. Cole, a little uncertain as to routine now, glanced up at the bridge. De Long nodded, so Cole piped down for mess, and with our ship pretty well on her beam ends, one watch laid below. There clinging to the stanchions, they ate the dinner which the imperturbable Ah Sam, still cooking in all that turmoil, had somehow, by lashing his pots down on the tilted galley range perhaps, managed to prepare.

At two bells, mess was over and most of us on deck again, hanging to the port rail. Soon we got another jam, listing us a little farther and still more raising our tilted bow, but the *Jeannette* took it well and I did not consider it anything serious, when suddenly, to everybody's alarm, my machinist Lee, whose station at the time was down on the fire-room floor running the little distiller boiler, shot out the machinery hatch to the deck, shouting:

"We're sinking! The ice is coming through the side!"

"Pipe down there, Lee!" ordered the captain sharply. "Don't go screaming that way to all hands like a scared old woman. You're an experienced seaman; if you've got any report to make, make it to me as if you were one! Come up here!"

Lee, white and shaking, climbed up the bridge ladder, his wound-weakened hips threatening to collapse under him. The captain beckoned me, then faced Lee.

"What is it now, Lee?"

"Her seams are opening below, sir! The sides are giving way!"

"Is that all?" asked De Long briskly. "What are you frightened at then? Here, Melville; lay below with him and find out what's wrong!"

With the reluctant Lee following, I climbed down into the fire-room. There was no water there.

"What in hell's the matter with you, Lee?" I asked angrily. "Do you want to shame me and the whole black gang for cowards? What set you off?"

"Look there, chief!" cried the agitated machinist. He led me into the starboard side bunker. We were well below the water-line. The air there was so full of flying coal-dust it was difficult to breathe, and as the ship thumped against the ice outside, new clouds of dust continuously rose from our panting sides. "Look at that! She's going fast!" yelled Lee, indicating with his torch. I looked.

The closely-fitted seams in the thick layer of planking forming our inner skin had sprung apart an inch or more, and as we watched, these cracks opened and closed like an accordion with startling frequency; but outboard of that layer we had a double thickness of heavy planking which constituted our outer shell, and though I could see traces of oakum squeezing out of the seams there, that outer planking, pressed by the ice hard in against the massive timbers of our ribs and trusses, was holding beautifully and there were no leaks.

"Keep your head next time, Lee," I advised gruffly as I came out of the bunker. "We're doing fine! Now mind that distiller, and don't salt up the water!" Blinking my eyes rapidly to clear them of coal-dust, I climbed on deck to inform the captain that there was no cause for alarm—yet.

So we lay for the next two hours, with the poor *Jeannette* groaning and panting like a woman in labour as the pack worked on her. At six bells, the captain, confident now that the worst was over and that she would pull through, took sudden thought of the future. The ship was a remarkable sight; what a picture she would make to print in the *Herald* on her return.

"Melville," said the captain, puffing calmly away at his meerschaum, "take the camera out on the ice and see what you can get in the way of a photograph."

"Aye, aye, sir!"

Early in the voyage, when Collins had been relieved of that task, I had become official photographer. I went to the dark-room, got out the camera, tripod, black hood, and a few of the plates which I myself had brought along and for which I had a developer. Stepping from our badly listed starboard rail directly on to the ice there, I picked a spot about fifty yards off on the starboard bow and set up my clumsy rig.

The view was marvellous. Heeling now 23° to starboard, the spar deck, covered with men clinging to the rigging, the rails, and the davits to keep from sliding into the scuppers, showed up clearly; while with her black hull standing sharply out against the white pack, and with bow and bowsprit pointing high in air and stern almost buried, the *Jeannette* looked like a vessel lifting while she rolled to a huge ice-wave. Never again would I see a ship like that!

I exposed a plate, then, for insurance, another; and folding up my rig, stumbled back over the ice to the ship, laid below to the dark-room on the berth deck, poured out my chemicals and proceeded with much difficulty (because of the extraordinary list) to develop the plates, which in that climate had to be done immediately or they would spoil. In the vague red light of a bull's-eye lantern, I was struggling in the dark-room with this job, when the ship got a tremendous squeeze, the berth deck buckled up under my feet, and amidst the roar of cracking timbers, I heard Jack Cole's shout:

"All Hands! Stations for Abandon Ship!"

Leaving the plates in the solution but extinguishing the red lantern, I hastily closed the dark-room door and ran on deck.

"Water coming up now in all the holds! I think that last push tore the keel out of her!" announced the captain briefly as I ran by him toward the cabin to get out the chronometers and the compasses. "I'm afraid she's through at last, chief!"

Behind me as I ran, I heard in rapid succession the orders to lower away the boats, to push overboard the sledges, and to commence passing out on the ice our emergency store of pemmican. Carefully I lifted out the two chronometers and the four small compasses which it was my job to save. Below me I could hear water gushing up into the afterhold, while from above on the poop deck came the creaking of frozen cordage and blocks as the falls ran out and our heavy boats dropped to the ice. As tenderly as I could, I gripped the chronometers, sprang out on the ice, and deposited my burden in the first cutter, already hauled a little clear of the ship's side.

The next few minutes, against a background of rushing water, screeching ice, and crunching timbers, were a blur of heaving over the side and dragging well clear our pemmican, sledges, boats, and supplies. Lieutenant Chipp, so sick in his berth that he could not stand, was dressed by Danenhower, and then the two invalids went together over the side, the half-blinded navigator carrying the executive officer who guided him.

I got up my knapsack from my state-room, tossed it into the cabin in the poop, and then turned to on our buckled deck in getting overboard our stores while below us the ship was flooding fast. De Long himself checking the provisions as they went over the side, looked anxiously round the spar deck, then asked sharply of the bosun:

"Where's the lime-juice?"

Our last cask of lime-juice, only one-third full now, was nowhere in sight.

"Down in the forehold, sor," said Cole briefly.

The forehold? Hopeless to get at anything there; the forehold was already flooded. De Long's face fell. There would be no distilled water any more; no more vegetables at all; nothing but pemmican to eat and salty snow from the floes to drink on our retreat over the ice, a bad combination for scurvy. The solitary anti-scorbutic we could carry was that lime-juice. He had to have it.

"Get it up!" ordered De Long savagely.

"I'll try, sor," answered the bosun dubiously. He went forward accompanied by several seamen, peered down from the spar deck into that hold. Water was already pouring in a torrent from the forehold hatch, cascading away over the berth deck into the lee scuppers. It was impossible to get into the hold except by swimming down against the current through a narrow crack left in the hatch opening on the high port side which, the ship being so badly listed to starboard, was still exposed. Yet even if a man got through, what could he do in the blackness of the swirling water in that flooded hold to find and break out the one right cask among dozens of others submerged there? But then that barrel, being only partly full, might be floating on the surface on the high side to port. There was a slight chance. Jack Cole looked round at the rough seamen about him.

"Any of yez a foine swimmer?" he asked, none too hopefully, for aside from the danger in this case, sailors are notoriously poor swimmers.

"I try vot I can do maybe, bosun." A man stepped forth, huge Starr, our Russian seaman (his name probably a contraction of Starovski), the biggest man on board. "Gif me a line."

Swiftly Cole threw a bow-line round Starr's waist. No use giving him a light; the water pouring through would extinguish it. He would have to grope in blackness. Starr dropped down to the berth deck. Standing in the water on the low side of the hatch, he stooped, with a shove of his powerful legs pushed himself through against the current, and vanished with a splash into the flooded hold. Cole started to pay out line.

How Starr, swimming in ice water in that Stygian hole amidst all sorts of floating wreckage there, ever hoped to find that one barrel, I don't know. But he did know that the ship, flooded far above the point at which she should normally sink, was held up only by the ice, and that if for an instant, the pack should suddenly relax its

Q

grip, she would plunge like a stone, and while the others on the spar deck might escape, he, trapped in the hold, would go with her.

With a thumping heart, Jack Cole "fished" the line on Starr, paying out, taking in, as the unseen swimmer fumbled amongst the flotsam in the black hold. Then to his astonishment, the lime-juice cask popped up through the hatch and following it, blowing like a whale, came Starr! Another instant and Starr, tossing the barrel up like a toy, was back on the spar deck, where all coming aft, Jack Cole proudly presented his dripping seaman and the precious cask on his shoulder, to the captain.

De Long, with his ship sinking under him, paused a moment to shake Starr's hand.

"A brave act, Starr, and a very valuable one. I'm proud of you! I'll not forget it. Now, bosun, get Starr here a stiff drink of whisky from those medical stores on the ice to thaw him out!"

The lime-juice, still borne by Starr, went over the side, the last of our provisions. The floes round about the *Jeannette* were littered with boats, sledges, stores, and an endless variety of everything else we could pitch overboard. With our supplies gone, I tried to get down again on the berth deck aft to my state-room to salvage my private possessions, but I was too late. The water was rising rapidly there, and was already half-way up the ward-room ladder, so I went back into the cabin in the poop above, where I had before tossed my knapsack, to retrieve it and get overboard myself.

The deck of the cabin was a mess of the personal belongings of all the ward-room officers—clothes, papers, guns, instruments, bear skins, stuffed gulls, that heavy walrus head over which Sharvell had once been so concerned (and apparently now, rightly) and Heaven knows what else. Pawing over the conglomerate heap was Newcomb, uncertain as to what he should try to save. As I retreated upwards into the cabin before the water rising on the ward-room ladder, De Long stepped into the cabin also from his upper deck state-room, and seeing only Newcomb fumbling over the enormous pile of articles, inquired casually:

"Mr. Newcomb, is this all *your* stuff?"

Pert as ever in spite of his illness, Newcomb replied with the only statement from him that ever made me grin:

"No, sir; it's only part of it!"

And even the captain, broken-hearted over his ship, looking at that vast heap, stopped to laugh at that.

But from the way the deck was acting beneath me, there was little time for mirth, so I seized my knapsack and walking more on the bulkhead than on the deck, got outside the poop, followed by the captain carrying some private papers, and Newcomb lugging only a shot-gun.

Things moved rapidly now on the doomed *Jeannette*. The ship started to lay far over on her beam ends, water rose to the starboard rail, the smoke-stack broke off at its base, hanging only by the guys; and then the ship, given another squeeze by the crowding ice, collapsed finally, with her crumpled deck bulging slowly upward, her timbers snapping, and the men in the port watch who were trying to snatch a last meal forward from scraps in the galley, finding their escape up the companionway ladder cut off by suddenly rising water, pouring like flies out through the forecastle ventilator to slide immediately overboard on to the ice.

It was no longer possible, even on hands and knees, to stay on that fearfully listed deck. Clinging to the shrouds, De Long ordered Cole to hoist a service ensign to the mizzen truck, and then with a last look upward over the almost vertical deck to see that all had cleared her, he waved his cap to the flag aloft, cried chokingly:

"Good-bye, old ship!" and leaped from the rigging to the ice.

Flooded, stove in, and buckled up, the *Jeannette* was a wreck. The pack had conquered her at last. Only that death grip with which the floes still clung tenaciously to her kept her afloat. With heavy hearts we turned our backs on the remains of that valiant ship, our home and our shield from peril for two long years, and looked instead southward where five hundred miles away across that terrible pack and the Arctic Sea lay the north coast of Siberia and possible safety—if we could ever get there.

CHAPTER XXVII

OUR situation was now truly desperate. There we were, thirty-three men cast away on drifting ice-floes, our whereabouts and our fate, whether yet alive or long since dead, totally unknown to the world we had left two years before, completely beyond the reach of any possible relief expedition. Five hundred miles away at the Lena Delta lay the nearest shore, where from the charts in our possession, we might expect to fall in on that frozen coast with native huts and villages such as long before we had visited at Cape Serdze Kamen, and find even a slight pretence of food and shelter and perhaps a little aid in getting over the next thousand miles south into Siberia itself to civilisation at Yakutsk.

How much of that five hundred miles before us was ice and how much was water, nobody knew. That a part of it, just north of Siberia, was likely to be water in the summer-time was certain, so we must drag our boats with us across the pack between us and that open sea or else ultimately, unable to cross it, perish when we came to the fringe of the ice-pack. A few uninhabited islands, the New Siberian Archipelago, lay half-way along the route, but we could expect no aid there of any kind nor any food. Grimly ironic on our Russian charts was the notice that all visitors were prohibited from landing on the New Siberian Islands unless they brought with them their own food, since the last party permitted a few years before to go there as fossil ivory-hunters had all starved to death for lack of game.

But at the Lena Delta, the charts showed permanent settlements and a book we had of Dr. Petermann's described in considerable detail the villages and mode of life there. Magazine articles published and taken with us just before we left San Francisco indicated that the Russian Government was then in 1879 about to open the Lena River for trading steamers from its mouth to Yakutsk, a thousand miles inland. Since it was now the middle of 1881, that

232

should be completed and the river steamers running.

So "On to the Lena Delta!" became our object in life, and to the Delta we looked forward as our Promised Land. But getting there seemed next to impossible. We were well acquainted with all previous Arctic expeditions. Not one in that long and tragic history stretching back three centuries, when disaster struck their ships, had ever faced a journey over the polar pack back to safety half so long as what faced us, and some on far shorter marches over the ice had perished to the last man!

Gloomily we faced our situation. We would have to drag our boats; we would have to drag our food; we were handicapped by one half-blind officer, by another too weak to stand, by several men, Alexey and Kuehne mainly, who, thrown on the ice, promptly had had a severe relapse of cramps from lead poisoning, and by the knowledge that many others of the crew were weakened by it and might break down at any time.

But worst of all we had to face was the pack itself. The most wretched season of the year for travelling over it was thrust upon us. Under the bright sun, the snow was too soft now to bear our sledges on its crust, but the temperature, from 10° to 25° F., was too low to melt it and clear it from our path. And as for the rough pack-ice itself, I knew best of all what travelling over that meant—the twelve-mile journey to Henrietta Island with a far lighter load per man and dog, had nearly finished me and my five men. And here, how many hundreds of miles of such ice we had to cross, God alone knew!

The loads we had to drag across the ice if we were to survive were enough to stagger the stoutest hearts. To carry our party over the open water when we reached it, required three boats, and the three boats weighed four tons. And to keep ourselves alive over the minimum time in which we could hope with any luck at all to reach Siberia, sixty days, required three and a half tons of food. Seven and a half tons at least of total dead-weight to be dragged over broken Arctic ice on a journey as long as the distance from New York half-way to Chicago! And the dragging to be done practically altogether (for at most our twenty-three remaining dogs could be expected to drag only one heavy sledge out of our total of eight) by men as beasts of burden—before that prospect we all but wilted.

But it was drag or die, and George Washington De Long was determined that not one man should die if he had to kill him to prevent it. For over De Long from the moment we were thrown on the ice had come a hardness and a determination which were new to us. Gone was the gravely courteous scholar, interested mainly in scientific discovery, scrupulously anxious to hurt no one's feelings if it could be avoided. The Arctic ice had literally folded up his scientific expedition beneath his feet, closing the books at 77° North, in his eyes practically a complete failure. That part was all over, gone with the ship, and with it vanished the scientist and the explorer whom we thought we knew. In his place, facing the wilderness of ice about us, stood now a strange naval officer with but a single purpose in his soul—the fierce determination to get his men over that ice back to the Lena Delta regardless of their hardships, regardless of their sufferings, to keep them on their feet tugging at those inhuman burdens even when they preferred to lie down in the snow and die in peace.

For five days after the crushing of our ship, we camped on the floes near-by, sorting out stores, loading sledges, distributing clothes, and incidentally nursing the sick. What he should take along was in the forefront of every man's mind. We had salvaged far more of everything than we could possibly drag—what should be left behind? De Long abruptly settled the question with an order limiting what was to be taken to three boats, sixty days' food, the ship's papers and records, navigating outfits, and the clothes each man wore, including his sleeping-bag and his knapsack, the contents of which were strictly prescribed. All else, regardless of personal value or desirability, must be thrown away. That was particularly trying to the men in freezing weather greedily eyeing the huge pile of furs, clothes, and blankets tossed aside to be abandoned on the ice, but there was the order—wear what you pleased in fur or cloth, trade what you had for anything in the pile if it pleased you more, but when you were dressed in the clothes of your choice, you left all else behind. The solitary article excepted was fur boots or moccasins;

of these each of us could have three pairs, one on, one in his knapsack, and one in his sleeping-bag along with his (half only) blanket.

But with the exception of much grumbling over the clothing to be left behind, there was no need of orders to enforce among the crew at least the abandonment of other weights; all the grumbling there was over what the captain ordered taken. Improvident as ever, the seamen growled over dragging so much pemmican, growled over dragging lime-juice, growled most of all over dragging the books and records of the expedition. But they didn't growl in the captain's presence. In range now of the steely glitter of those hard blue eyes, strangely new to them, they only jumped to obey. Still, among themselves (and I was always with them now) there was a continual growl over the loads building up on the sledges, and as for what articles they were themselves to carry, I saw seamen weighing sheath-knife against jack-knife to determine which was lighter, and then tossing the heavier one away.

The start of our life on the ice the night we lost our ship was inauspicious. Dead tired from superhuman labouring, first in hurriedly getting stores off the ship and then in dragging them over the ice to what looked like a safe floe two hundred yards away from her, we turned in at midnight, camping in five small tents, five or six men and an officer in each, stretched out in a row on a common rubber macintosh. At one a.m., with a loud bang the floe beneath us split, the crack running right through De Long's tent, and the ice promptly opened up. Had it not been for the weight of the sleepers on the ends of the mackintosh there, the men sleeping in the middle would immediately have been dumped down the crevice into the sea! Even so, practically helpless in their sleeping-bags, they were rescued with difficulty, while all the rest of us, weary as we were, hastily turned out to move our whole camp across the widening crack to another floe. By two a.m., this was done, and again we turned in, leaving only Kuehne on watch. At four a.m., as he was calling Bartlett, his relief, from my tent, he announced suddenly:

"Turn out if you want to see the last of the *Jeannette!* There she goes, there she goes!"

I leaped up and out of the tent. There was the listed *Jeannette* coming slowly upright over the pack, for all the world like a ghost rising from a snowy tomb. The floes holding the ship, as if satisfied at having fully crushed the life from her, were evidently backing away. She came erect, her spars rattling and creaking dismally as she rose, then the ice opening further, she started to sink, with accelerating speed. Quickly the black hull disappeared, then her yards banged down on the ice, stripped from the masts, and in another instant, over the fore-topmast, the last bit of her I ever saw, the dark waters closed and the sturdy *Jeannette* had sunk, gone to an ocean grave beneath the Arctic floes!

I stood a moment in the cold air (the temperature had dropped to 10° F.) with bared head, a silent mourner, but thankful that we had not gone with her, then crawled sadly back to my sleeping-bag.

On the sixth day after the crushing of the ship, our goods were sorted, our sledges all packed, and we were ready to go southward. Eight sledges, heavily loaded with our three boats and our provisions, carried about a ton each; a ninth sledge, more lightly loaded carrying our lime-juice, our whisky, and our medical stores, was considered the hospital sledge; while a tenth, carrying only a small dinghy for temporary work in ferrying over leads, completed our cavalcade.

To minimise the glare of the ice and the strain of working under a brilliant (but not a hot sun) all our travelling was to be done at night when the midnight sun was low in the heavens, with our camping and sleeping during the day when the sun being higher, his more direct rays might be better counted on to dry out our soaked clothing.

At six p.m. on June 17, we started, course due south.

Our first day's journey was a heart-wrenching nightmare which no man there was likely to forget till his dying gasp. The dogs were unable to drag even their one sledge; it took six men in addition to keep it moving. And so bad was the snow through which the sledges sank and floundered, that we found it took our entire force heaving together against their canvas harnesses to advance the boats and their sledges one at a time against the snow banking up under the bows of the clumsy boats.

Dunbar had gone ahead, planting four black flags at intervals to mark the path which he as pilot had selected for us to follow, the fourth and last flag, only a mile and a half along from the start, being the end of our first night's journey. But so terrible was the going that by morning only one boat, the first cutter, had reached that last flag; the runners had collapsed under three of the sledges, stalling them; a wide lead had unexpectedly opened up in a floe half-way down our road, blocking the other sledges and requiring them to be unloaded and ferried over it; Chipp (who, with Alexey and Kuehne, in spite of being the sick were dragging the hospital sledge) had fainted dead away in the snow; Lauterbach and Lee had both collapsed in their harnesses, Lauterbach with cramps in his stomach, Lee with cramps in his legs; and by six a.m., when our night's journey should have been finished and all hands at the last flag pitching camp, we had instead broken down and blocked sledges scattered over that mile and a half of pack ice from one end to the other!

It was sickening. Twelve hours of man-killing effort and we had made good over the ice not even one and a half miles!

Willy-nilly, we made camp, breakfasted, and turned in at eight a.m., for our exhausted men could do no more without a rest. But for Surgeon Ambler, there was no rest. While the remainder of us, dead to the world, slumbered that day, Ambler, who as much as anyone the night before had toiled with the sledges along that heart-breaking road, laboured over the sick, struggling to get them on their feet again for what faced them that evening.

That night we turned to once more, repairing runners, shifting loads, digging sledges out of snowbanks in which they were buried, and fighting desperately to advance all to the first camp. Regardless of a temperature of 20° F., we perspired as if in the tropics, and tossing aside our parkas, worked in our under-shirts in the snow. All that night and the next night also, we laboured thus. By the second morning following, thank God, we had all our boats and sledges together there, and tumbled again into our sleeping-bags, wearied mortals if ever there were such on this earth! Three nights of hell to make a mile and a half of progress! It was worse even than my journey to Henrietta Island had led me to believe could be possible.

And then that day, of all things in the Arctic, it started to *rain!* Miserable completely, we sat or lay in our leaking tents, soaked, muscle-weary, and frozen, while the cold rain trickled over us and over the icy floors of our tents. But while I had thought no creatures could possibly be suffering greater misery than we, I changed my mind when I saw our dogs, cowering in the rain, snuggle against our tent doors, begging to be admitted to such poor shelter as we had. So soon, with men and beasts shivering all together, the picture of our misery was completed.

In the midst of all this, Starr opening up the rations for our midnight dinner, found in a coffee can a note addressed to the captain, which he brought to him. It read:

This is to express my best wishes for success in your great undertaking. Hope when you peruse these lines you will be thinking of the comfortable homes you left behind you for the purpose of aiding science. If you can make it convenient drop me a line. My address,

G. J. K——,

Box 10, New York City.

"*Apropos,* eh, Melville? I guess we're thinking of those homes, all right," commented De Long bitterly, showing me the note. "Where's the nearest post-box so I can drop that imbecile a line?" Near-by was a crack in the floe. "Ah, right here." De Long scribbled his initials on the note, drew an arrow pointing to the writer's address, and dropped it into the crack. "Now we'll see how good the sea-going Arctic mail service is. At that, it may get to New York before we do," added the captain grimly.

Before we moved off from this camp, the captain decided to check the loads to make sure we were taking nothing more than was absolutely necessary. The first thing he discovered was that flouting his order about clothing, Collins had smuggled into our baggage and was taking along an extra fur coat. Immediately, under the captain's angry eye, it went flying out on the ice. And the next thing he found was that in their knapsacks (which were towed along stowed inside the whale-boat) the seamen almost without exception were taking some small mementoes of the cruise, trifling in weight in themselves, in the aggregate under our circumstances, a considerable burden. They went sliding out on the ice alongside Collins's coat.

And then having cleaned house, the captain waited for the rain to end.

Since it rained all night, we stayed in camp, getting a rest, if such it can be called. Next night we were under way again on a new schedule, the load of supplies on each sledge now cut in half (except of course for the boat sledges) the idea being to lighten up our overloaded sledges so we could move them to the designated point more easily and with less danger of breaking runners, then unload and send them back empty for the other half of their cargoes. Working this way we started out, only to find half a mile along a crack in the ice, not wide enough for a ferry, too wide to jump with the sledges. Here the ice broke up with some of our sledges floating off on an island, stopping all progress till we had lassoed some smaller cakes for ferries and on these we rode over our remaining loads, finishing our night's work with hardly half a mile gained and everybody knocked out again.

So for the next four days we struggled along, sometimes making a mile a day; once, by great good luck, a mile and a quarter. The going got worse. Pools of water from the late rain gathered beneath the crusts of snow and thin refrozen ice. As we came along, the surfaces broke beneath us, leaving us to flounder to our knees through slush and ice water. More sledge runners broke; Nindemann and Sweetman were kept busy at all hours repairing them. Chipp got worse, Alexey vomited at the slightest provocation, Lauterbach looked ready to die, and Lee staggered along on his weakened legs as if they were about to part company at his damaged hips. Danenhower, of all those sick, while he could hardly see, at least had some strength, and was added to the hospital sledge to help pull it under Chipp's pilotage. Ahead Dunbar scouted and marked out our road south by compass, then with a pickaxe endeavoured to clear interfering hummocks from that path, aided a little in that by Newcomb. I bossed the sledge gangs and kept them moving, putting my shoulders beneath a boat or a sledge when necessary to get it started. Ambler, when not tending his patients, armed with another pickaxe, helped Dunbar clear the chosen road. And bringing up the rear was De Long, supervising the loading, checking food issues, and relentlessly driving us all along.

On June 25, we had been under way eight days since starting south. By such gruelling labour over that pack as men cannot ordinarily be driven to, even to save their own lives, and which in this case only the overpowering will of De Long rendered possible, we had made good to the southward by my most liberal calculation a total distance over the ice of five and one-half miles. I contemplated the result with a leaden heart. Even should the ice extend southward only one hundred miles out of the five hundred we had to cover (which seemed far too good to be true), at that rate of advance it would take us one hundred and fifty days to cross to open water. Long before that, unless we died of exhaustion first as now looked very probable, our sixty days' rations would have been consumed and we should be left to perish of starvation midway of the pack.

In despair, I gazed at our three cumbersome boats, overhanging at both ends their heavy sledges, the last of which after soul-wrenching efforts my party had just dragged over rough hummocks into camp. Around it the men, too exhausted even to go to their tents, were leaning their weary bodies for a moment's rest before they undertook the labour of lifting again their aching and frozen feet. Those massive boats, like millstones round our necks, were what were killing our chances. With our food alone, divided into reasonable sledge loads, we might make speed enough to escape, but with those boats——! Incapable of division, the smallest over a ton in weight, the largest over a ton and a half, dragging those boats was like dragging huge anchors over the floes. If only we could abandon them! But with a sigh, I gave up that dream. With an open sea somewhere ahead, the boats were as necessary to us as the pemmican. But only five and a half miles made good in our first week, when we were strongest! It looked hopeless. We could only labour onward and pray for a miracle. I quit thinking and turned toward my tent, my supper, and my sleeping-bag.

On the way, I bumped into Mr. Dunbar, just returning from a preliminary scouting trip over our next night's route. Dunbar, hardly fifty, hale and hearty, a fine example of a seasoned Yankee skipper when first he joined us, now with his face wrinkled and worn, looked like a wizened old man staggering under a burden of eighty years at least, and ready to drop in his tracks at the slightest provocation.

"Well, captain," I sang out jocularly to cheer him up a bit, "what's the good word from the front? Sighted that open water we're looking for yet?"

The whaler looked at me with dulled eyes, then to my astonishment broke down and sobbed on my shoulder like a baby. I put a fur-clad arm gently round his heaving waist to comfort him.

"What is it, old shipmate? Can't you stand my jokes either?"

"Chief," he sobbed, "ye know it ain't that; I like everything about ye. But that ice ahead of us! It's terrifically wild and broken, and so chock-full o' holes, chief, I could hardly crawl across! We'll never get our sledges over it!" The weeping old seaman sagged down in my arms, his grey head nestling in my beard.

"Don't be so sure, mate," I said with a cheeriness I didn't feel. "My lads are getting so expert heaving sledges over hummocks, I'm thinking of putting 'em on as a flying trapeze act in Barnum's Circus when we get back, and making us all as rich as Commodore Vanderbilt in one season! Come on, captain, forget it; let's have a cup of coffee to warm us up—no, let's belay the coffee. Come to think of it, I guess I still got drag enough with Dr. Ambler to work him for a shot of whisky apiece for a couple of good old salts like us." And I led him away to the hospital tent, where Ambler, after one look at Dunbar, hardly needed the wink from me to produce without a word his medical whisky.

Leaving Dunbar with the doctor after swallowing a drink myself,

I started again for my own tent, but once more I was stopped, by the captain this time, who beckoned me to join him in the snow alongside the deserted whale-boat. All hands were in their tents by now, working on their cold pemmican.

"What have you made our mileage to the south so far, chief?" opened the skipper listlessly.

"Being generous, about five and a half miles, sir." I looked at him puzzled. The captain knew our progress, logged daily in his journal, even better than I. Surely he wasn't keeping me from my supper just for that.

De Long nodded, continued:

"That's right, over the ice of course. Melville, I'm sorry to say that to-day I got some good sights of the sun for the first time since we started. Chief," and his voice broke as he looked at me, "the north-west drift has got us! We're twenty-five miles farther *north* to-night than the day we started!"

CHAPTER XXVIII

TWENTY-FIVE miles farther north than when we started! Coming on top of all else, that was a knock-out blow. With sagging knees, I leaned against the gunwale of the whale-boat, looked at the haggard captain, asked faintly:

"Sure?"

"Yes," he replied mournfully, "sure. Two meridian altitudes of the sun and a couple of Sumner lines and they all check. I couldn't believe the first sight myself; I thought my sextant was bad. But after checking that with nothing wrong, I spent all day getting check sights; you saw me shooting some of them. No question about it now. But, for God's sake, don't tell anyone! Not even another officer. If the men knew, I couldn't get 'em to lay a hand on another sledge. They'd just sit down here and wait to die!"

I nodded at that. Who could blame them? Caught on a treadmill, why should they torture themselves with such labour as slaves would lie down under and suffer themselves to be lashed to death rather than rise and endure, when the only result was that they were being carried backward five times as fast as their puny efforts pushed them forward? If the men learned the results of those disastrous sights, we were finished! But weren't we finished anyway, whether they knew it or not?

"That's bad, captain," I mumbled, my brain numbed at the news. "What can we do now?"

"I haven't figured it out fully yet, chief; I want your advice on what I have in mind. That's the only reason I told you; that, and maybe the thought that at least one other officer ought to know where we are and where we're going in case anything happens to me. Since Chipp and Danenhower, both my deck-officers, are knocked out, you're the only officer left I can talk to. But this mustn't get out! It'd kill Chipp, who's in a bad way anyhow, like hitting him with an axe. And what it'd do to the men, you can guess!" He finished with a ghastly smile.

Vaguely I felt that that might be a mercy to Chipp; indeed it might well be a mercy to all of us should a kindly Providence then and there somehow brain us all with an axe and end our sufferings. But of that I said nothing. The captain had hinted at something further to be done. What, I wondered, was it? Abandon the boats, speed up our progress, trusting to luck the ice held out under us till we got to the New Siberian Islands, there to live (or starve rather) on moss and Arctic willow till perhaps some year a chance party of ivory hunters landed and rescued us? A thin chance, that! I looked curiously at De Long.

"Well, skipper, no thanks for telling me. I could've got more work out of my gang if I were still in blissful ignorance like the others than I can now, but I'll keep on doing my best. And if my advice is any good, you're welcome to it. What's your idea?"

"I know that well enough, chief. I would gladly have suffered under that knowledge alone, but the safety of the whole party requires someone else to know, and unfortunately to suffer with me. But no use talking about that now. Let's get along with what's next." He jerked out from beneath his parka a chart, unfolded it, spread it on the midship thwart of the whale-boat. On it, marked by a bold cross enclosed in a red-ink circle, was a spot in latitude 77° 18′ N. where the *Jeannette* had sunk, a spot unfortunately for us now, far to the south-east of the small pencilled cross in latitude 77° 43′ N. which marked our present position as shown by the captain's latest sights.

"There's just one possible thing we can do, Melville; change the course from due south as we're heading now, to south-west. The ice is drifting clearly enough north-west. It's like crossing a river current; no use bucking into it as we're doing now going south, even if land is closest that way. The only hope is to cut across perpendicular to the current and ultimately you get to the other bank, even though the current carries you downstream meanwhile. The same with us, we've got to cut directly across the drift, that is, south-west; and some day, regardless of how far north we're carried, we'll come to the edge of this pack and can launch our boats—provided we live long enough! And to insure that, we're going to go on shorter rations now to stretch out our food," he concluded significantly. "I've decided all that already, Melville. All I want to know from you—you're working the men—is how thin I can cut it and still let you keep on driving them, harder even than now!"

I thought a moment. Our daily rations, pemmican, hard bread, and coffee, were none too generous in quantity as it was for the terrific physical effort the men were labouring under. But who knows what men can stand? I was learning all the time.

"Well, skipper, cut a third off. That'll stretch it out for ninety days instead of sixty. I doubt that we can keep up long on such short rations, but maybe we can knock over a few seals to help out now and then. Anyhow, I'm willing to try."

"Good!" agreed De Long. *"Nil desperandum!* I knew I could count fully on you, chief; I always can. We'll start that programme, new course and all, to-night." He started to fold up the chart again, when his eye fell on the pencilled cross marking our position. "77° 43' North," he muttered. "Farthest North for us yet. And for ever too, on this cruise, anyway, I hope. Say, Melville, you remember that silk banner you used in claiming Henrietta Island? Well, it's in the first cutter, alongside the whale-boat here. There's something else about that flag my wife made for me. She wanted me to fly it in celebration when we made our 'Farthest North.' Let's hope we're celebrating that glad event right here and now. But the crew knows about that too, I think. If I fly *that* flag over this camp, they'll smell a rat right off." He looked furtively round. All our men were apparently still in the tents. No one was in sight. "I think I can take a chance to please Mrs. De Long on this; it's little enough I've done for her since she married a sailor. Here, Melville, lend a hand."

Together we drew from its case and unwrapped that silken banner, a moderately large American ensign beautifully embroidered round the edges by Emma De Long's loving hands for her husband's ship. And stealthily, keeping it below the gunwales of the first cutter lest someone looking from a tent should see and wonder, we fully extended it horizontally. Then standing on the ice alongside that open boat, with De Long to starboard and me to port, we two, looking certain death on the pack in the face, waved that banner beneath the Arctic sky in latitude 77° 43′ from the Equator—Farthest North for George Washington De Long, if not for the *Jeannette!*

And then, leaning over the gunwale, De Long buried his weather-beaten face in the rustling silk folds of his wife's flag, kissing it fervently, while I, clinging to the other side, closed my wet eyes in silence. Well did I remember that June day in 1879, almost exactly two years before in San Francisco, when at our commissioning Emma De Long, a lovely figure, had herself proudly manned the halliards and hoisted that banner to the masthead on the *Jeannette!* But the *Jeannette* was gone beneath the floes, and far away at that moment, I envisioned Emma De Long, a different woman now, worn by two years of hourly dread over news that never came, praying for the safety of the man before me, who with heaving shoulders was caressing his country's flag, the solitary symbol of his wife's love still left in his possession.

De Long straightened up.

"All right, Melville. I think my wife'll be glad to know when it gets back we flew her flag at our Farthest North. Come on, let's fold it up; carefully now, so nobody'll know we had it out and perhaps guess why."

Silently I obeyed. We rolled up the flag, slid it into its oilskin case, carefully re-stowed the case as before. The chances of our ever getting back were slight now, but as I shoved that case under the thwarts in the cutter, bachelor as I was, I hoped that even though our own bodies might soon be stretched in death over that desolate ice-pack, somehow that flag in the boat might survive the drift, some day to be picked up and returned to the one person who would sense among its silken folds the message that it bore.

CHAPTER XXIX

WHEN we got under way that night on June 25, we headed south-west instead of south. Burning in my breast were De Long's words *Nil desperandum!* and my faith in him from his faith in me, rose. The ice to the south-west, thank God, Dunbar reported as not so bad as he had found it to the south, though Heaven knows how that could have been, for we had to bridge and ferry five times in one mile, and in many places to get our sledges over inescapable hummocks blocking our path, we had to build inclined planes of snow to their crests and other inclines down the lee slopes, then heave our sledges up one slope like Egyptian slaves building the Pyramids, and brace ourselves back to ease them down the other side. We couldn't even coast down the lee slopes, for then the sledges buried their noses so deeply in the banks at the bottom that extricating them was horrible work.

So like horses (though sometimes sea-horses as we ploughed water to our waists) we worked along through the ensuing week, making about a mile and a half a day over the ice. We were all in a bad way from exhaustion, and oddly enough the brilliant sun, cold though it was around 28° F., burned and blistered our faces and added to our general suffering. On top of that a mental trouble became noticeable; the men were grumbling because no news of our position or of our progress had been posted, for they had all seen the captain taking sights and felt that the results ought to be made known. As the week drew along and nothing was said, they began to get suspicious, but none dared ask the captain; when they questioned me, I merely shrugged my shoulders, saying:

"Don't ask me, boys. I'm only an engineer! Why should I bother with the navigation?"

As for the growling, which was plentiful over the shortened rations, I could point out that our progress was slower than originally expected, so we must naturally stint ourselves to stretch them

247

out longer, and thus allayed any suspicion on that score.

But De Long had a busy time dodging his other officers, lest they ask embarrassing questions. With Chipp and Danenhower this was not difficult, for Chipp could hardly walk and Danenhower could hardly see. Keeping away from them was easy. Ducking Dunbar was much harder, but since the ice-pilot was ahead laying out the road most of the night, the skipper with some finesse managed to steer the discussions into safer channels on the few occasions when he couldn't avoid him. Ambler, however, turned out to be a Tartar who from the very nature of his duties the captain couldn't keep away from. Finally, concluding that with Ambler confidence was better than suspicion, he acquainted him also with the reasons for our sudden change of course, and I must say for the doctor that I think he took it better than I did, for early in the week when he was told, he was more than having his hands full between swinging a pickaxe on the roads and tending his patients, especially Chipp.

Lieutenant Chipp, carried from his sick-bed when the ship sank, was in a bad way from exposure and sleeping on the ice, despite the fact that he was the only person allowed to take an extra coat. Even the week's rest before we started sledging south helped him little and he fainted in his harness the first day out. After that, though hauling no longer on the hospital sledge, he had since barely managed to stagger along with it as it went.

The day we started south-west, so badly off was Chipp that slow as we went that day, he could not hobble well enough to keep his emaciated and pain-racked body up with the hospital sledge and was delaying even its snail-like progress. De Long, bringing up the rear-guard, ordered him to climb aboard the sledge and ride. Chipp made no move to get aboard, but instead staggered onward. Without a word, De Long picked him off the ice, laid him gently on his back on the sledge, and ordered briefly:

"You stay there, Chipp, or I'll hand you a court-martial for insubordination! You're delaying our progress when you walk!"

Poor Chipp, broken-hearted at being made a burden for his overladen shipmates to drag, tried to roll off the sledge to the snow, but so weak was he that he could do no more than turn on his face when he stuck, clawing feebly, trying to pull himself off the sledge. Failing even in that, he looked pitifully up at De Long by his side.

"Don't make them drag me, captain, please! I'm all gone anyway. Take me off!" he begged. "You'd better leave me behind right here!"

"Shut up, Chipp," ordered De Long abruptly. "You'll do what I say like everybody else in this outfit! Quit worrying; you're going to get better. But better or worse, nobody gets left behind while there's a man alive able to drag him along!"

He motioned to Lauterbach, Alexey, and Danenhower, dragging the sledge.

"Get under way now, men. I'll help you till I can send someone else." And with the captain pushing and the others pulling, the sledge started again with the enfeebled Chipp face down on the load, scarcely able to cling to the lashings, weeping bitterly.

We got along. But I might here mention that for every mile of progress over the ice that we made, we had to walk thirteen miles. To advance a boat sledge took the whole working party together; to advance a provision sledge half-loaded took half the party so to get our three boats and four provision sledges (the dogs handled the other one) along one mile meant that seven times under load and six times empty-handed, thirteen times in all, the staggering working party had to traverse that mile of ice. If the edge of the pack should by God's grace turn out to be no farther than a hundred miles from our starting point, still when we reached it we should have tramped thirteen hundred miles over that terrible ice, seven hundred of those miles dragging inhuman loads! If it were twice as far—God help us then!

So we went along, over what Jack Cole, ruefully tugging in the lead harness, called:

"'Tis the rocky road to Dublin, my byes. Yo heave! Shure an' we should be nearly there by mornin'!" But I knew it would be many a morning yet, if ever, ere Jack raised Dublin or anything like it over that ice horizon.

July 3 arrived with good enough weather for the captain to get another set of observations of the sun, which on working out, he communicated to me. The new position was in latitude 77° 31′ N., longitude 150° 41′ E., which was to some degree gratifying, for while it was still thirteen miles north of where we left the Jeannette, it was thirteen miles generally south-west of where we were on June 25, and checked very well with both the course we had been steering and our distance logged over the ice since then, twelve miles. This was cheering, for it seemed to indicate no ice drift at all for the last eight days, and things began to look up. Only thirteen miles more and we should be as far south as when we started sixteen days before! Naturally, while all this cheered De Long, Ambler and myself, the knowledge would have cheered nobody else, so no notice of it was posted and no mention made.

July 4 we celebrated on the ice, without any fireworks or speeches, simply breaking out our small boat flags (the woollen ones only) and, so to speak, dressing ship. De Long was excessively blue all day, for it was the third anniversary of the day in Le Havre, France, when Miss Bennett had christened his ship with her name, *Jeannette,* and he had listened to many glowing speeches of what was expected of her. Looking at the three small boats which were all that was left of his command made De Long decidedly sick. Had there been only the safety of himself to consider the day his ship went down, I am sure De Long would have gone down with her.

By way of a feast in honour of the day we had our usual short allowance of cold pemmican which we ate thankfully. I may say here that pemmican (which is a mixture made of beef pounded more or less to a powder, mixed with raisins, and then the whole stirred up in boiling fat which when cold is packed in cans) while a highly-nutritious and palatable food served in cold slices, which we ate like cake, as a steady diet gets infernally tiresome. Alexey, on this day, with a naïve faith in the white man's powers, feeling that a holiday called for something better, in all seriousness told the doctor that he would take mutton instead!

At this, Lee, my machinist, who was also in the doctor's tent, very gravely informed the doctor of the best way to make Rhode Island clam chowder, which he felt was the only proper dish for any July 4th banquet, and the poor doctor, with all this gastronomical advice bringing back recollections of past Independence Day feasts back in old Virginia garnished with everything from savoury baked hams to candied sweet potatoes, found his mouth watering, so that he lost all interest in his cold pemmican and fled from the tent.

Under way again that evening, we stumbled along as before, heaving, holding back, building ice bridges, ferrying on bobbing floes across the water leads when we did not fall into them. De Long, his mind a little relieved about the drift, spent fewer hours in the rear and most of the night tramping over our route, for the first time beginning to take some notice of individuals and what they were doing. Coming up to one bridging job, where I had a piece of floating ice jammed into a crack some fifteen feet wide while the crew were dragging sledges across it to the southern side, he noticed that Collins, standing at the edge of the gap, was holding the line securing the make-shift bridge in place.

"Mr. Collins," said the captain icily, "you have many times in disrespectful language informed me that you didn't ship to be treated as a seaman. I can't allow you to go home, claiming that I forced you to work as one even to save your own life. Give that rope to one of the men!"

Collins made no move to obey. Instead, for perhaps half a minute he stood glaring like a tiger at De Long, till the latter, noting Seaman Dressler close alongside, sternly ordered:

"Mr. Collins, give that rope to Dressler, and don't let me catch you putting your hand to another line until I order you! You are still under suspension awaiting trial and don't you forget it!"

Collins, ready to burst with anger, slowly passed over the line and without a word dropped to the rear.

We moved along. Under the continued burning rays of the sun, the snow melted and drained away from the surface, making the going a little easier, and our consumption of food lightened up our loads, still further aiding our speed, but our personnel troubles increased.

Ambler was particularly burdened. Ten days of riding on the

sledge and careful medical attention had so built up Chipp that he could walk again, and with that little improvement, he began to nag the doctor to put him off the sick-list and restore him to duty in command of one of the parties. Danenhower also, his physique improved by the enforced exercise he was getting in walking after his long confinement aboard ship, began to make the same demand though he could hardly see through his one heavily-shielded eye. Ambler naturally enough refused both requests. As a result, daily when he came into his tent after having wielded a pickaxe all night long on the roads, and crawled horribly tired into his sleeping-bag to rest, it was only to listen to his two blessed invalids exchanging sneering remarks about his medical competence because he would not restore them to duty. Finally, unable to stand it further, he burst out:

"For God's sake, shut up, both of you! Dan, if you'd obeyed my orders on the ship, one of your eyes would be well now! And you, Chipp, a little while ago were begging us to leave you on the ice to die! Now that you're both barely able to get yourselves along, you want *me* to risk other men's lives by putting them in your charge, and I'll be damned if I will! Was ever a doctor cursed by two such patients!"

But if his patients aggravated him, his helpers on the road work tried his very soul. In charge of the road-building gang, Ambler had as assistant labourers Lee and Newcomb, and to draw along the sledge with the dinghy which was assigned to him for working in the open leads, seven of such miserable, broken-down dogs that they were worthless for any work on the heavy sledges and only an irritation on his lighter one. But even so, Ambler might philosophically have accepted the situation and kept on as he was, doing most of the work with a pickaxe himself, had it not been for Newcomb. For both the broken-down dogs and Lee with his shaky legs were at least doing their poor best. Quite to the contrary, our naturalist, though fully recovered from his indisposition, infuriated the doctor, himself manfully swinging a pick, by the piddling efforts which he was please to pass off as work. Patiently Ambler showed him how to swing a pick on the hummocks; then getting no results from him, sharply ordered him to turn to, only to find Newcomb more interested in pertly answering back than in obeying. For two days Ambler, with his southern temper slowly rising, stood it, merely remarking grimly to me one night:

"If that Yankee chatterbox doesn't soon do some work instead of answering me back every time I speak to him, he's going to get some medical attention that'll astonish him!"

I watched them working a little ahead of me that night as I trod back and forth with the sledges. We had only two pickaxes in our whole outfit. Newcomb, at the base of a steep hummock, was using one axe in a pretence of picking at it, while I could see Ambler, standing on its slippery crest, nervously tightening his calloused hands about our other pickaxe handle as if debating whether to swing then on Newcomb's head or wait till he was a little surer of his footing. While the doctor was in this uncertain frame of mind, Newcomb below him quit picking at the ice altogether, lashed a line to his pickaxe to make it serve as an anchor for something or other, and then, sad to relate, overboard into an open water lead went the precious pickaxe, line and all, a total loss!

In spite of the real tragedy which the loss of that pickaxe meant to us, what happened next made me roar. Ambler's fingers closed firmly on the handle of our sole remaining axe, apparently determined, poor as his shot now was, to swing and make an end of the gadfly below him; then changing his mind, he leaped from the hummock, stopped only a second to wave the pick in Newcomb's face while he bellowed:

"You bird-stuffing idiot! If I weren't afraid of breaking our last axe on your worthless skull, I'd kill you with it!" and dropping the pick on the floe, he ran off to find the captain.

A few minutes later he was back, all smiles, to find Newcomb, instead of manning the last axe, casually scanning the sky for something really important—gulls perhaps.

"It's all fixed now, Mr. Newcomb," said Ambler cheerfully. "I offered the captain to trade you for another worthless cur for my sledge, but he couldn't find one poor enough to make it a fair exchange, so bless his heart, he gave me Seaman Johnson for my road gang and said I could do what I pleased about you. So now you're fired! Get out of here!"

Newcomb, pert as ever, enquired:

"Discharged, I presume you mean to infer? How welcome! What am I to do now?"

"Tie your shot-gun round your neck for ballast and jump after that pickaxe, if you want to do me a favour!" advised the doctor, fingering the last pick significantly. "Now get out of my way while I work, or there may yet happen what will go down in the log as a most regrettable accident!"

CHAPTER XXX

DE LONG was left with the problem of how to make Collins and Newcomb useful members of our primitive community. While Collins, before the captain noticed him, had done useful work when it suited him to help, I have little doubt that it was only for the Machiavellian purpose of building up a brutal mistreatment case against the captain. Had he been ordered to work steadily in harness like the others, he would either have flatly balked or else have done it only as a martyr, neither of which situations the captain was prepared to cope with. Newcomb's case was a little different. Had he been my problem, I am confident that the toe of my boot, properly applied a few times, would have startlingly unchange his outlook both on work and on keeping his mouth shut when spoken to, but De Long was constitutionally opposed to physical persuasion. Casting his eye about the floes, the skipper observed that seals were again occasionally in evidence, and decided that since both our Indians, Alexey and Aneguin, were labouring like all the others as pack-horses, he might well substitute Collins and Newcomb for them as hunters. Hunting being in all civilised circles a gentleman's privilege, neither of these pseudo-seamen officers could well maintain that it was beneath the stations for which they had shipped, and if they shot anything, it would be of real value in stretching out our precious food supply, let alone giving us a change from the pemmican which constituted the fish, flesh, fowl and vegetable of our unvaried menu.

So promptly providing Collins with a rifle and Newcomb with the shot-gun which he had carried from the wreck, both were turned to on the floes to see what they could do in earning their passage, while our straggling line of boats and sledges moved on over the pack.

But even so Collins was not satisfied. A member of the party messing and sleeping in the captain's tent, his main business in life seemed to be sizing up what he could find wrong with De Long's management of the retreat, to add in his private notes to whatever else had had accumulated in the way of (in his eyes) errors in the captain's judgment. But he must have had a tough time of it, for the only thing that apparently displeased him now was as he related it to his *confidant,* Bartlett:

"The skipper's always too infernally polite to me, seeing that I'm served before he helps himself to pemmican, and making sure my place in the tent's all right before he'll crawl into his sleeping-bag."

And under these embarrassing conditions, Collins began his life as a hunter.

July 10, far to the south-west, Dunbar sighted a faint cloud which he announced as land, gravely assuring us that the New Siberian Islands were in sight. While the skipper was dubious of its being land at all, knowing (what Dunbar and most of the party did not) that the nearest charted land, those New Siberian Islands, were still a hundred and twenty miles off, so that what was seen was either a mirage or a new discovery, the effect on our progress was magical —over none too good ice we made three and a quarter miles that day! In the clearer atmosphere, the skipper got some sights for the first time in a week which when worked out placed us in latitude 77° 8′ N., longitude 151° 38′ E., to our joy showing that at last we were south of 77° 18′ N. from which we had started. But both the skipper and I stared in amazement at our new position, for having by dead reckoning and compass made sixteen miles to the south-*west* over the pack during the week, the sights showed that our actual change of position in those seven days was twenty-seven miles to the south-*east.* So once again the current had us, carrying us where it would, but since this time it was increasing our southing, we could only be thoroughly grateful.

That night when (still keeping our actual position a secret) it was announced that not only were we doing well over the ice, but that a southerly drift was helping us along, there was a roaring cheer as the straining men in harness leaned forward, and we got the boats away in grand style.

For two weeks we struggled on to the south-west, sometimes cerain we saw land, sometimes certain we didn't. But it was discouraging work. Fog, snow, and hail made our lives miserable, and between the everlasting ferrying over open leads and the ploughing through pools of surface slush, we kept our clothes continuously soaked in ice water. Aside from the discomfort of stretching out in wet clothes to sleep in a wet bag on a rubber sheet sunk in a puddle on wet ice, the interminable wetness began to finish our moccasin-and boot-soles, and now there wasn't a tight pair left in the ship's company. These soles, made of "oog-joog" skin, a rawhide from a species of seal, were fine when dry in ordinary snow or ice, but when wet, they softened to resemble tripe and then under the strain of men heaving hard against sharp ice with their feet to drag the sledges, they soon let go. As long as the spare "oog-joog" brought from the *Jeannette* held out, we patched away till all hands stood on a mass of patches as they worked, but when it gave out (and it very soon did), we were in a bad way for substitutes. First we tried leather, stripping it from the oar-looms, but leather was not only too hard and slippery for use on the ice, but our supply didn't last long, and we were quickly reduced to canvas, to sennet-mats woven to hemp-rope by the seamen, to rag-mats, and even to wooden soles carved from what little planking our carpenter could strip out of the bottom boards in our boats. None were in any degree satisfactory—one hard heave on sharp ice would often tear the soles of a man's boots—and frequently before the end of a night's hauling I would have half a dozen men straining at the sledges with their bare feet on the ice, even their socks completely worn through, while the rest of the gang, whose soles still clung on, would be spurting a mixture of slush and water from their torn moccasins at each step.

Between the lodestone effect of the dim land ahead of us, less snow, a little smoother ice, and lighter sledges, we speeded up. The ice improved to the point where we could drag a boat with only half our party, thus advancing two loads at once and having to tramp only seven miles for each mile made, instead of thirteen as before. But the cracks in the ice increased in frequency and ferrying and bridging over them made our lives a nightmare, the mental strain of for ever riding heavy sledges over bobbing ice-cake, which threatened to capsize each instant, being indescribable. And to add to our worries, our dogs began to get fits, four of our best ones spinning dizzily in their harnesses before dropping on the ice, frothing at the mouth when we cut them out of the traces.

One pleasing incident occurred amidst all our hardships. After ten days of hunting, Collins finally shot a seal in an open lead, which prize was handsomely recovered by Ambler and Johnson in the dinghy before it sank. For this we were doubly thankful—after using his grease to tighten up our leaking boots, we dined most luxuriously on stewed seal, fried seal, and if only we had had an oven, we might have had roast seal. But he went very well as it was; after a month on cold pemmican, it was a feast long to be remembered!

July 16 we struck tough going. The ferrying grew worse than ever; Erichsen crossing a lead capsized with his sledge and we lost three hundred pounds of pemmican, a serious blow. A few minutes later, trying to get to a high hummock to inspect the distant land now more visible ahead, De Long tried to jump a wide lead, the ice broke under him and he went in up to his neck. He might well have gone completely and for ever had not Dunbar, who was with him, at that point grabbed him by what he thought was his fur hood, but which was actually his whiskers, and nearly jerked the skipper's head off pulling him out by his moustaches!

Finally, on this day, the doctor discharged Chipp from the sick list, though doubtful as to how long it might be before Chipp broke down again. This resulted in a shuffle in commands—Chipp relieved me in charge of the working force; I relieved Ambler in charge of the road gang; and Ambler with only Danenhower left as a regular patient, was detailed to work with Dunbar in scouting out the road. The doctor offered to join the sledge gang in harness, but we were doing better there, so the skipper refused. He preferred to use Ambler simply as scout and medical officer, hoping that his terribly calloused, corned, and chapped hands might recover enough for proper surgical work should an accident make any necessary.

The skipper worked out some sights. The latitude, reliable, was 76° 41′ N.—twenty-eight miles gained to the south in six days—fine progress, much more than we were logging over the ice. The longitude, doubtful, put us at 153° 30′ E., indicating we were still going south-east though we were heading south-west, but we were not greatly concerned over that. Anything to the south was cause for gratitude.

We dragged along five days more. Newcomb at last shot something, a gull he called a *mollemokki,* interesting ornithologically to him, perhaps, worthless to us for food, certainly. The ice grew rotten; we had more trouble with it. Our men, their eyes and minds affected by the ice, easily deluded by mirages, were now seeing land in nearly all directions, south of us, west of us and even *north* of us! And not a day went by when someone didn't see open water ahead of us, fine wide-open sea in which we could launch our boats, toss away our sledges, and sail homeward in comfort!

Instead of that we soon bumped into the worst mess of ice we had yet encountered, a jumble of small lumps and water, with numberless large floes tipped on end vertically. With my road gang and our solitary pickaxe, I started the herculean job of clearing away some of these hummocks so we might proceed, and was busily at it when the doctor, bless his soul, came in to report that by retracing our path northward half a mile, we could then go due west till we got on the flank of that broken ice, after which we might go southwest again. I snapped at that; the job ahead of me was like tunnelling through a mountain with a tooth-pick. So back over our trail we went with our boats and sledges.

Getting across even that better path was a heart-rending job, for the rotten floes would hardly stick alongside each other, till finally using all the lines we had, like Alpine climbers we lashed the floes together while we crossed over, seriously hampered by a dense fog. It was a long stretch. In the middle of it, we came to morning, our usual time for piping down to camp and rest during the day, but the captain, seriously alarmed at the prospect of that rotten and moving ice distintegrating under us while we slept, belayed the usual camp. So without rest and only a brief stop for supper, we kept on, till after twenty-three hours of terrific labour we came in the late afternoon to a solider floe and stopped at last to rest our weary bones.

The captain, feeling rightly enough that what we now most sorely needed was sleep rather than cold supper, gave the order for all hands to turn in. This the men in my tent thankfully did and were soon stretched out in their sleeping-bags, but in the next tent, assigned to Danenhower, Newcomb and five seamen, Newcomb immediately sounded off.

"This is a fine way," he said sarcastically, "to treat men who have been working so hard; ordering them to turn in without anything to eat!"

Lieutenant Danenhower peered in surprise through his dark glasses at the naturalist, who had done nothing all day but carry a small shot-gun.

"Maybe it is hard for the men who are working, Newcomb," he said quietly, "but for you and me who haven't done a blessed thing, it isn't, and we shouldn't be the first to complain now."

Newcomb ran true to form. Instead of taking the hint thus delicately conveyed, he retorted angrily:

"I wasn't speaking to you; I was speaking to these men. I don't count myself in the same category with you. I'm a worker!"

Newcomb a worker! Danenhower could hardly believe his ears. But not wishing to start a row before the men, and not wanting anyone, least of all a man who passed as an officer with them, to encourage them in the belief that they were ill-treated, he ordered curtly:

"Pipe down, Newcomb! That's enough on that!"

But piping down was one thing not in Newcomb's psychology. Answering back suited him better.

"No, I won't!" he piped up. "I don't take orders from you. And now that the crisis has come, I'm going to meet the issue! You've made yourself disagreeable to me right along, but I'm an officer too, and it's got to stop!"

Had Dan been able to see in more than a vague blur, the issue would undoubtedly have met Newcomb's jaw then and there. As it was, without further words, Dan stumbled from the tent to report the still-spouting naturalist to the captain for endeavouring to foment trouble in the crew.

In five minutes, Newcomb was placed under arrest to be taken home for court-martial on two charges:

I. Using langauge tending to produce discontent among the men; and

II. When remonstrated with by Lieutenant Danenhower, using insolent and insubordinate language.

There being little further De Long could then do, he deprived Newcomb of his shot-gun, ordered him to keep in the rear as we proceeded, and sternly warned him meantime not to annoy anyone working.

So when late that night we got under way, we had two officers under arrest—the surly Collins, who seemed to spend much of his time unburdening his wrongs in the ears of my fireman Bartlett, but in between times making himself useful as a hunter, and Newcomb, who was thoroughly useless for anything.

The land which Dunbar weeks before had sighted across the ice, undoubtedly a newly-discovered island not on the charts, was now in plain sight only a few miles off, bearing westward. Through bad gales and over broken pack, with occasional floebergs suddenly shooting into the air near us, we worked toward it. July 23, the captain's sights showed no change in our latitude since the 16th; in that time between our own efforts and the erratic drift, we had been taken twenty-eight miles due west and were now fairly close to our new island. We struggled along toward it over badly-moving ice, but at least this ice was firm and many of the floes were large. Collins finally shot another seal, but it sank before the dinghy could get to it, and we sadly saw our visions of a second feast dissolve into cold water. Next morning we pitched camp as usual, with the land tantalising us not three miles off, but mostly hidden in fog. Soon after turning in, the man on watch shouted:

"Bear!" and instantly out of their sleeping-bags popped Alexey

s

and Aneguin, eager to get the first bear sighted since our ship sank. We heard a couple of shots, and our mouths began to water. Um-m! Bear steaks for dinner! But it was all wasted, for soon the two Indians were back, empty-handed and disgusted! The bear had been in such excellent trim that they had had to fire at a thousand yards on a rapidly-reciprocating target as that bear humped himself over the ice, and they had of course missed. However, it didn't matter much, claimed Alexey, as the bear was only a dirty brown one and not very big, a remark which prompted the captain to ask innocently:

"Sour grapes, Alexey?" but Alexey only looked at him puzzled. Grapes, sour or otherwise, never grew in his latitude, so I'm afraid he missed the point. Quieting our disappointed stomachs as best we could, once more we turned in.

But we got the bear. In the late afternoon, Seaman Görtz, who had the watch the while the rest of us slept, spotted him once again. This time Görtz kept his mouth shut while the bear advanced to within five hundred yards of our camp, and then, unnoticed, our look-out managed to crawl within a hundred yards of him to plant two bullets in that bear where they did the most good!

Now that we had him, he turned out to be a very fine bear indeed, even Alexey admitting that ungrudgingly, and soon the air over that floe was filled with an appetising aroma of sizzling bear-steaks that fairly intoxicated us. We envied no man on earth his evening meal that night as, disdaining pemmican, we gorged ourselves on bear. But we needed it. When we broke camp and started for the island ahead, we found ourselves with nothing but moving ice over which to work our sledges.

For two days, mostly in fog, we fought our way toward that island, with the floes breaking under us, sliding away from us, and the whole pack alive around us. A gale blew up, and on the off side of the hummocks about us a bad surf broke and kept us drenched. Finally, on the third day, we found ourselves opposite the dimly-visible western tip of the island, with nothing but a forlorn chance left of ever making the solid ground that so desperately we ached to rest ourselves on. With but a few hundred yards remaining before the pack finally drifted us past it for ever, we sighted ahead a long floe of heavy blue ice extending in toward the land, with only a few openings between the floe and ours. We bridged the gaps, bounced our sledges and boats over, and made good a mile and a half across that floe. There we found more broken ice and water, which with difficulty we started to cross in the fog by passing a line to a floe beyond and using a smaller cake as a ferry-boat, when suddenly the fog lifted and there over our heads, some 2,500 feet high, towered a huge cliff, and sweeping past it as in a mill-race were the floes on which we rode!

We finished our ferry, ending on a moderately-sized floe drifting rapidly past the fixed ice piled up at the base of the cliff, with the south-west cape, our last slim chance to make the land, not far off. For over two weeks we had dragged and struggled toward that island; now in despair we found ourselves being helplessly swept by it!

Our little floe, covered with sledges, men and dogs, whirled and eddied in the race, spinning crazily, and threatening to break up any moment, when we noted that if only it should make the next spin in the right direction, it might touch a corner against the ice fringing the land. We waited breathlessly. It did!

"Away, Chipp!" shouted De Long, and in an instant our sledges started to move off that spinning floe. The first got away perfectly, the second nearly went overboard, the third sledge shot into the sea, carrying Cole with it, and the fourth was only saved by Erichsen who, with superhuman strength, shoved an ice-cake in for a bridge. We couldn't get the boat sledges over; our floe was already starting to crack up. Working frenziedly as it broke, the few of us left on the floe pushed the boats, their sledges still under them, off into the water and the men already landed started to haul the boats over to them, when away drifted the last remnant of that ice-cake, carrying with it De Long, Iversen, Aneguin and me, together with six dogs! For a few minutes we were in a bad way, threatening to drift clear of the island on that tiny ice-cake with no food, except perhaps the dogs; while the men ashore ran wildly along the ice-foot, unable to help us in any manner.

Fortunately for us, a little farther along a swirl drove our floe in against a grounded berg for a second and dogs and all, we made a wild leap for it; successfully too, for only three of us landed in the water. Aneguin, the Indian, proved the best broad jumper. He landed safely enough on the berg and dragged the rest of us up and out.

Soon reunited again, behind our dripping captain the entire ship's company straggled across the ice-foot to solid ground (the steep face of the cliff), where clinging to the precipice with one hand, the captain for the third time on our voyage displayed his silken banner, proudly rammed its staff for a moment into the soil, and exclaimed:

"Men, this is newly discovered land. I therefore take possession of it in the name of the President of the United States, and name it:

"Bennett Island!"

The men, most of them (except the five who had landed with me on Henrietta Island) with their feet on solid ground for the first time in two years, cheered lustily. Jack Cole then sang out:

"All hands, now. Three cheers fer Cap'n De Long!" in which all again joined except Collins and (for the first time in his life managing to keep his mouth shut when anybody gave him an order) Newcomb.

But our happy captain, not noticing that, turned to his executive officer and jocularly remarked:

"We've been a long time afloat, Mr. Chipp. You may now give the men all the shore leave they wish on American soil!"

It was July 28 when we landed; we stayed a week on Bennett Island, resting mainly, while Nindemann and Sweetman worked strenuously repairing our boats. All were badly damaged and unseaworthy from the pounding they had received in the pack. The whale-boat especially, our longest boat, had suffered severely and every plank in its stern was sprung wide open. Sweetman did the best he could in hurrying repairs, pouring grease into the leaking seams and refastening planks, but it was a slow job, nevertheless.

While this was going on, the men explored Bennett Island, which we found to be of considerable extent (we never got to its northwest cape), probably thirty miles long and over ten miles wide, very mountainous, with many glaciers, running streams, no game we ever saw, and thousands of birds nesting on the cliffs. This island, at least three times the size of Henrietta Island, nicely finished off the honours due the Bennett family, for we now had one each for Mr. Bennett, his sister, and his mother.

Geologically, we found the island interesting. I discovered a thick vein of bituminous coal, and Dr. Ambler found many deposits of amethyst crystals, but what took our fancy most were the birds. We knocked down innumerable murres with stones, which, fried in bear's grease, we ate with great relish. But they proved too much for Dr. Ambler's stomach, laying him in his tent for over a day.

On August 4, with the boats all repaired, we made ready to leave. To the southward of Bennett Island, the pack looked to us badly broken up with enough large water openings to make it seem that thereafter we could proceed mostly in the boats among drifting floes, keeping the sledges for use when required. To this end, since the dogs would be less necessary and feeding them on our pemmican an unwarranted further drain on our stores, De Long ordered ten broken-down dogs to be shot to avoid their suffering should we abandon them, keeping only the twelve best for future sledging, including husky Snoozer, who was by now quite the captain's pet.

By sledge over the pack we had travelled almost exactly a hundred miles in a straight line from where the *Jeannette* had sunk to Bennett Island, though over the winding track as we actually crossed the drifting ice we had dragged our sledges more than a hundred and eighty miles, and in so doing had ourselves tramped far beyond a thousand miles on foot. We prepared hopefully to rely from then on mainly on our boats, and for this purpose the captain rearranged the parties, breaking up the sledge and tent groups in which we previously had journeyed.

Into the first cutter with himself he took a total of thirteen—Dr. Ambler, Mr. Collins, Nindemann, Erichsen, Kaack, Boyd, Alexey, Lee, Noros, Dressler, Görtz and Iversen.

Into the second cutter (a smaller boat), under Lieutenant Chipp's command, he put ten—Mr. Dunbar, Sweetman, Sharvell, Kuehne, Starr, Manson (later transferred to my boat), Warren, Johnson, and

Ah Sam (who later to lighten still further the second cutter, was transferred to De Long's boat).

Into the whale-boat, of which he gave me the command, also went ten—Lieutenant Danenhower, Mr. Newcomb, Cole, Bartlett, Aneguin, Wilson, Lauterbach, Leach, and Tong Sing.

Thus we made ready, with De Long commanding the largest and roomiest boat, Chipp commanding the smallest boat, and me in command of the whale-boat, considerably our longest craft, though not our greatest in carrying capacity. And promptly there flared up in the Arctic an echo of that Line and Staff officer controversy agitating our Navy at home. (At home, it lasted until the Spanish War showed that we engineers were as important in winning battles as deck officers, and maybe more so.)

I, as an engineer officer, belonged to the Staff; Danenhower, as a deck officer, belonged to the Line, which alone maintained the claim to actual command of vessels afloat. A whale-boat was not much of a vessel, but nevertheless Danenhower, when he heard of the assignments, promptly informed me he was going to protest to the captain.

"Go ahead, Dan," I said. "That's perfectly all right with me." So the navigator went to the captain to object to a staff officer being given command while he, a line officer, was put under my orders. In that congested camp on Bennett Island, he didn't have far to go to find the skipper.

"Captain," asked Dan, "what's my status in the whale-boat?"

"You are on the sick-list, sir," replied De Long.

"Who has command of the boat?" persisted Dan.

"Mr. Melville, under my general command."

"And in case of a separation of the boats?" questioned the navigator. "Suppose we lose you?"

"In that case," said the captain, "Mr. Melville has my written orders to command that boat and what to do with her."

"Am I under his orders?"

"Yes, so far as it may be necessary for you to receive orders from him."

"But that puts me under the orders of a staff officer!" objected Dan strenuously.

"Well, you're unfit to take command of the boat yourself," pointed out the skipper. "You can't see, Mr. Danenhower. I can't put you on duty now. So long as you remain on the sick-list, you will be assigned to no military control whatever."

"Why can't I be put in a boat with a line officer, then?" asked Dan, the idea of having to report to a staff officer rankling badly.

"Because I have no line officer left to put in that boat with you, and because I have seen fit so to distribute our party. I want one line officer in each boat. In an emergency, Mr. Melville may wish to have your advice on matters of seamanship."

"Well then," replied Danenhower vehemently, "I remonstrate against being kept on the sick-list."

"But you're sick and that's nonsensical," said De Long curtly.

"Why, sir, haven't I the right to remonstrate?"

"You have, and I've heard you, and your remonstrance has no effect," replied the captain bluntly. "I've had the anxiety of your care and preservation for two years and your coming to me on these points now is simply an annoyance. I will not assign you to duty till you're fit for it, and that will be when the doctor discharges you from the sick-list. I will not put other people's lives in jeopardy by committing them to your charge, and I consider your urging me to do so is very un-officer-like conduct."

Taken aback at this barbed comment on his complaint, Danenhower asked hesitantly:

"Am I to take that as a private reprimand?"

"You can take it any way you please, Mr. Danenhower," concluded the irritated De Long, walking away to supervise the loading of the first cutter, leaving the crestfallen navigator no alternative but to come back to join me in the whale-boat.

"Don't take it so hard, Dan," I suggested. "Too bad about your eyes, of course, but it can't be helped now. We've always been the best of friends and we're not going to let this change things. As long as you're in the whale-boat, you can count on me, old man, not to say or do anything that'll hurt your feelings as an officer. Hop in, now; we're shoving off!"

Already delayed two days by bad weather, on August 6 we got

away from Bennett Island, with intense satisfaction, though the wind had died away, being able to get under way in our boats under oars, carrying the sledges and our twelve remaining dogs. The boats, of course, packed with men, food, records, sledges, and dogs, were heavily overloaded and in no condition to stand rough weather, but we had smooth water and we made two miles before bringing up against a large ice island. Here we lost most of our dogs, who not liking water anyway, and objecting still more to the unavoidable mauling they were getting in the crowded boats from the swinging oars, promptly deserted the moment they saw ice again, by leaping out on the floe, and we were unable to catch them. We worked around the ice-pack in the boats, by evening getting to its southern side, where we camped on the ice, with five miles between us and Bennett Island, a good day's work and a heaven-sent relief from sledging.

The weather was startling clear. Looking back, we got a marvellous view of the island. When we had first reached it in late July, its appearance was quite summery with mossy slopes and running streams, but now winter had hit it with a vengeance. Everything on it was snow-covered and the streams were freezing. We regarded it with foreboding. The first week of August and the brief Arctic summer was fading away, with four hundred miles before us still to go on our journey to the Lena Delta. We must hurry, or the open leads we now had for the boats would all soon freeze over.

For two weeks we stood on to the south-west, boating and sledging. With luck in pushing away the ice with boat-hooks we might make five or six miles between broken floes before we met a pack we could not get through afloat, when it was a case of unload the boats, mount them on their sledges, and drag across the ice. By the second day of this, we were down to two dogs, Snoozer and Kasmatka, all the rest having deserted, but these two special favourites were kept tied and so prevented from decamping. The boat work, whether under sail or oars, was hard labour. There was no open sea, merely leads in the open pack, and over most of these leads, the weather was now cold enough to freeze ice a quarter of an inch thick overnight. We found we could not row through this, so the leading boat, usually the first cutter, had to break a way, and all day long men were poised in her bows with boat-hooks and oars breaking up the ice ahead. And we had before us several hundred miles of this!

The weather was bad, mostly fog, snow squalls, and some gales, but because of the vast amount of floating ice, there was no room for a heavy sea to kick up, and when a moderate sea rose, we always hauled out on the nearest floe. And so camping on the ice at night, hauling out for dinner, and making what we could under sail or oars in between when we were not sledging over the pack, we stood on to the south-west for the New Siberian Islands. At the end of one week's journeying, the snowfalls became frequent and heavy, troubling us greatly, though they did provide us with good drinking water which was an improvement over the semi-salted snow we got on the main pack. By now it was the middle of August, sixty days since we had left the *Jeannette,* and the expiration of the period for which originally we had provided food. We were hundreds of miles from our destination, and our food was getting low. Of course, had it not been for our going on short rations soon after our start, our position would now be precarious, since the few seals, birds, and the solitary bear we got, while luxurious breaks in our menu so long as they lasted (which wasn't long) meant little in the way of quantity.

By August 16, nine days under way from Bennett Island, we had made only forty miles—not very encouraging. Next day we did better—ten miles under sail with only one break, but the day after, it was once more all pulling with the oars and smashing ice ahead and slow work again. But on August 19 we saw so much open water that we joyfully imagined we were near the open sea at last. We loosed our sails and until noon went swiftly onward with the intention of getting dinner in our boats for the first time without hauling out on the ice, and then continuing on all night also. Suddenly astern of us we saw Chipp's boat hastily douse sail, run in against a floe, and promptly start to unload.

There was nothing for the rest of us (cursing fluently at the delay) to do except to round to and secure to the ice till Chipp came up, and long before he had managed that, the ice came down on us from all sides before a north-east wind, so that shortly it looked as if there was nothing but ice in the world. Chipp finally sledged his boat over the pack to join us and we learned the ice had closed on him suddenly, stove a bad hole in his port bow, and he had to haul out hurriedly to keep from sinking.

By three p.m., Chipp had his boat repaired, using a piece of pemmican-can for a patch, and we were again ready. Each boat had its sledge, a heavy oaken affair, slung athwartships across the gunwales just forward of the mast. Abaft that, the boats were jammed with men and supplies, the result being that they were both badly overloaded and top-heavy.

With great difficulty we poled our way through ice-drifts packed about us to more open water and made sail again before a freshening breeze, De Long in the first cutter leading, my whale-boat in the middle, and Chipp with the second cutter astern of all. We felt we must be nearing the northerly coasts of the New Siberian Islands, which we hoped to sight any moment and perhaps even reach by night.

The breeze grew stronger and the sea started to kick up. My whale-boat began to roll badly, taking in water over the gunwales, and at the tiller I found it difficult to hold her steady on the course, though with some bailing we got along fairly well, and so it seemed to me did the first cutter ahead. But the second cutter astern, the shortest boat of the three, was behaving very badly in that sea— rolling heavily, sticking her nose into the waves instead of rising to them, and evidently making considerable water. Hauling away a little on my quarter and drawing up so he could hail the captain ahead, Chipp bellowed down the wind:

"Captain! I've either got to haul out on the ice or heave overboard this sledge! If I don't, I'll swamp!"

De Long decided to haul out. He waved to Chipp and me (he being to leeward, we couldn't hear him) indicating that we were to haul out on a floe near-by on our lee side. The near side of that floe, its windward side, had a bad surf breaking over the ice, so we tried to weather a point on the floe and get around to its lee where we could see a safe cove to haul out in, but our unwieldy boats would not sail close enough to the wind, and we failed to make it. Chipp's case by this time was desperate; his boat was badly flooded, and in spite of all the bailing his men could do, the waterlogged cutter seemed ready to sink under him. There was nothing for it but to land on the weather side of the ice, which dangerous manœuvre, with a rolling sea breaking badly on the floe and shooting surf high into the air, was skilfully accomplished without, to our intense relief, smashing all our boats beyond repair.

The gale grew worse. It was now 7.30 p.m. and beginning to get dark. (Between the later season of the year and our being farther south, we no longer had the midnight sun with us, but instead about eight hours of darkness.) There was no hope of further progress that night, so we pitched camp on the floe, while the gale started to push ice in about us from all directions.

That night before supper, the captain called Chipp and me to his tent. The question for discussion was the boat sledges. We had since leaving Bennett Island broken up all our other provision sledges and burned them for fuel. Chipp strenuously insisted that the boat sledges be treated likewise immediately.

"Captain," he said, "I'm surprised I'm here to talk about it even! My boat's so top-heavy with that sledge across her rails, a dozen times I thought she'd either founder or capsize. And a man can't swim a minute in these clothes in that ice water. If she'd sunk under me, long before you or the chief could've beat back against that wind to pick us up, we'd all be gone!"

With Chipp's facts, honestly enough stated, De Long was inclined to agree and so was I, but the question was too serious to be decided out of hand. On our first journey across the pack, the sledges were our salvation, and it was the heavy boats (holding us back like anchors) which we then gravely considered abandoning lest our party perish before we ever reached water. Now the situation was reversed; it was those boats, dragged across the ice at the cost of indescribable agony, which had become our main hope of escape, but still could we afford to abandon the sledges which so obviously now imperilled our safety in the boats? We were not yet out of the pack; one had only to poke his head through the tent-flap to see as much ice as ever we had seen. And if we had to sledge over much more of the pack to get south, without those boat sledges we couldn't do it. What then should be done with the sledges?

With our lives very likely depending on that decision, we considered it deeply. The conclusion, concurred in by all, was that the certainty of disaster if we kept the sledges, out-weighed the possibility of being now caught permanently in pack ice, unable to move except by sledging, and De Long finally gave the order to burn the sledges. In a few minutes, knives and hatchets in the hands of sailors eager to make an end of those incubi before the captain could change his mind, had reduced them to kindling and they were burning merrily beneath our pots. No man regretted seeing them go who had toiled in the harnesses dragging them and their bulky burden of boats across the ice-pack, labouring as men have never done before, and as I hope may never have to again.

Further to help Chipp, the captain in expectation next morning of a long voyage among the New Siberian Islands, decided to even matters somewhat more by removing from Chipp's stubby cutter, only sixteen feet long, part of its load. Accordingly he decreased its crew by two men, taking Ah Sam (our Chinese cook who had since the sinking of the *Jeannette* with nothing but pemmican on the menu, not cooked a meal, serving instead only as a beast of burden like all others) into the first cutter with him, and sending Manson, a husky Swedish seaman, to join my crew in the whale-boat. In addition, De Long took into the first cutter part of Chipp's supply of pemmican, still more to lighten his boat, which was certainly a worse sea boat than either my twenty-five-foot-long whale-boat or his own twenty-foot-long cutter.

With these rearrangements, we camped for the night in the midst of a howling gale drifting snow about our tents, the while we earnestly hoped that the wind would break up the pack in the morning, and allow us to proceed.

But instead, for ten wearing days we lay in that camp, unable to launch a boat and unable, of course, to sledge them over the broken pack, while the weather varied between gales with heavy snow and dismal fogs, and we ate our hearts out in inaction, watching our scanty food supplies constantly melting away with no progress to show for it. Our hard bread gave out altogether, our coffee was all used up, and our menu came down to two items only, pemmican and tea three times a day, with an ounce of our fast-vanishing lime-juice for breakfast to ward off scurvy. To save what little alcohol we had left (we had been using it for fuel for making tea and coffee) we continued to burn up the kindling from our boat sledges, but long before we broke camp, even that was all gone, and we started again on our precious alcohol.

The tobacco gave out (each man had been permitted to take one pound with him from the *Jeannette*). To the captain, an inveterate pipe-smoker, this was a severe trial and left him perfectly wretched, till Erichsen, who still had a trifle left, generously shared with him the contents of his pouch. De Long declined to take more than a pipeful, but Erichsen insisting on an even division of his trifling remnant, the skipper found he had enough for three smokes. Immediately seeking out the doctor and Nindemann, he divided with them, and together they puffed on their pipes, in a mixed state of happiness and despair watching the last tobacco from the *Jeannette* curling upward in smoke wreaths into the Arctic air.

Next day, like the others, they were smoking used tea-leaves and getting little solace from them.

Our second day in this camp, through a rift in the fog we sighted land twenty miles to the southward, in the captain's opinion the island of New Siberia, one of the largest islands of the New Siberian group and the one farthest eastward in that archipelago.

As the days went by and in the fog and snow we drifted westward with the pack before an easterly gale, the knowledge of that unapproachable island added to our aggravations. We could do little except repair our boats (which, using pemmican tallow, rags, and lamp-wick for caulking materials, Nindemann and Sweetman laboured at) and wait for the pack to open, a constant watch night and day being set with orders that if a lead appeared, we should immediately launch our boats into it. But none showed up. In desperation at the delay, which was bringing us face to face with the prospect of starvation, De Long again sent for Chipp and me.

"Mr. Chipp," he asked, "can you move your boat across this ice to the land?"

"No," said Chipp flatly. "It'll stave in her bottom trying to ride her on her keel."

"Mr. Melville," turning to me, "can you get the whale-boat across? Is this any worse than when you landed with the dinghy on Henrietta Island?"

"Captain," I replied sadly, "no worse, but it's as bad; the ice is just as much alive. And I didn't take the little dinghy to Henrietta Island, even on her sledge; I left her at the edge of the moving pack. I can get the whale-boat across this ice to that island if you order me, sir, but when she gets there, she'll be worthless as a boat."

"Well, in that case," remarked De Long, bitterly disappointed at our views, "it's no use taking them there." And while he didn't voice it, there was little question but that he deeply regretted having ever cut up the boat sledges. In my opinion, however, sledges or no sledges, we couldn't safely get those boats through to the land over that swirling ice between. We started to leave.

"Hold on a moment," ordered De Long, pulling a book out from under his parka, "there's something else." He pushed his head out the tent-flap, called to the man on watch in the snow. "Send Seaman Starr in here!"

In a moment or two, Starr, with his snow-flecked bulk practically filling the tent, stood beside us. The captain opened the book. It was in German, one of Petermann's publications, the best we had on New Siberia and the Lena Delta. Starr, aside from his Russian, could also read German, and as he translated, De Long, Chipp, and I followed on the chart, putting down Petermann's data on the islands, and especially on the Lena Delta, where near Cape Barkin were marked winter huts and settlements, a signal station similar to a lighthouse, and the indication that there we could get native pilots to take us up the Lena. At this time the captain warned me that should we be separated, Cape Barkin was to be our rendezvous. At that point the delta formed a right-angled corner. To the westward from Cape Barkin, the coast ran due west; but at the cape itself, the coast turned and ran sharply south for over a hundred miles, while through both the northern and the eastern faces of this corner-like delta, the Lena discharged in many branches to the sea. But Cape Barkin at this corner we must make—there between the pilots and the settlements shown by Petermann, our voyage would end and our troubles would be over. The remainder of our journey home would merely be a tedious and probably a slow trip on reindeer sledges southward from the Siberian coast inland fifteen hundred miles to Irkutsk, then a long jaunt westward by post coaches to Moscow, and so back to America. The captain marked it all out, made two copies of his chart, one for me and one for Chipp, and then dismissing Starr, told us:

"There are your charts with the courses laid out to Cape Barkin. As I informed you in my written order at Bennett Island, Melville, if unfortunately we are separated, you will continue on till you make the mouth of the Lena River, and without delay ascend the Lena to a Russian settlement from which you can be forwarded with your party to a place of safety and easy access. Try to reach some settlement large enough to feed and shelter your men before thinking about waiting for me. And the same for you, Chipp. That's all, gentlemen. Be ready to start the instant the ice breaks." He drew out his pipe, ending the discussion. We took our charts and departed, leaving the skipper trying to light off a pipeful of damp tea-leaves.

On August 29, after ten days of fuming in idleness, during which time our pack drifted first westward and then southward, the weather cleared a bit and we found ourselves between Fadejovski and New Siberia Islands, and closer to Fadejovski, the western one of the pair. At noon, Dunbar, scouting on the pack, reported a lead half a mile away. Immediately we broke camp, and carrying our provisions on our backs while we carefully skidded our boats along on their keels, we dragged across that half-mile of floe to the water and launched our boats, thankful even for the chance the remainder of the afternoon to fight our way through swirling ice-cakes to the southward. The drift in that lead was rapid, the broken ice there was violently tumbling and eddying, and as we swept down the bleak coast five miles off Fadejovski Island unsheltered from the intense cold, with oars and boat-hooks savagely fending off those heaving floes on all sides of us to keep our frail boats from being crushed, it was like making passage through the very gates of hell! For two horrible days we worked along the coast fighting off impending death in that swirling maelstrom of ice, when with the pack thinning somewhat, we managed at last to work our way to land on the southerly end of Fadejovski, three weeks under way since leaving Bennett Island, and humbly grateful to find ourselves disembarking still alive.

We stayed one night on a mossy slope trying to thaw our frozen feet by tramping on something other than ice, and as Dunbar expressively put it: "Sanding our hoofs." They needed it. The most pleased member of our party was Snoozer, now our sole remaining dog, who joyously tore round chasing lemmings, while we sought for real game which we didn't find. And that night was served out our last ration of lime-juice which so heroically salvaged by Starr from the sinking *Jeannette,* had shielded us from scurvy for two and a half months on our tramp over the ice. But we saw the last drops of that unsavoury medicine disappear without regret and without foreboding for the future, for now we were nearing the open sea and our voyage was nearing its end.

Next morning we shoved off from the south end of Fadejovski, only to discover despondently that we had embarked on a twelve-days' Odyssey through the New Siberian Archipelago before which our previous sufferings seemed nothing. We had not wholly lost the ice; instead we had only added to our previous perils some new ones—vast hidden shoals, bitter freezing weather, long nights of sitting motionless and cramped in our open boats, while the Arctic winds mercilessly pierced our unshielded bodies, and the hourly dread of drowning in a gale.

It was seventy miles over the sea to Kotelnoi, the next island westward in the group. To get there, instead of being able to sail directly west, we found we had to stand far to the southward of Fadejovski to clear a shoal, getting out of sight of land. When night caught us far out in the open sea, we discovered even there shoals with less than two feet of water, over which a heavy surf was breaking badly. Standing off into deeper water, we beat all night into the wind to save ourselves from destruction, for we had no anchors in our boats. Wet, miserable, frozen by spray coming over, we stayed in the boats, so crowded we could not move our freezing legs. At dawn we stood on again westward, with streaming ice bobbing all about us, travelling before a fresh breeze all day. In the late afternoon, having lost sight of Chipp and the second cutter, his boat being unable to keep up, we finally spotted a floe sizeable enough to camp on. De Long signalled me to stop; we promptly secured to it and waited for Chipp to catch up, meanwhile for the first time in thirty-six hours stretching our wet forms out in our sleeping-bags on the ice, while a gale blew up, snow fell, and the sea got very rough, which gave us grave concern over our missing boat. By daylight there was so much pack-ice surrounding our two boats, it seemed unbelievable we had arrived there by water, and our anxiety for Chipp increased. We lay all day ice-bound, all night, and all next day, occasionally sighting the mountains on Kotelnoi Island, perhaps ten miles to the westward. And then Chipp and the second cutter finally showed up, coasting the north side of our floe, half a mile away across the pack, and soon Chipp and Kuehne, walking across the ice, were with us. They had had a terrible time the night we lost them; long before they sighted any floes, the gale caught them, and over the stern where Chipp and Dunbar sat steering, icy seas tumbled so badly that all hands bailing hardly kept the boat afloat till they finally found a drifting floe. When at last he steered in under the lee of the ice, but one man, Starr, was still able to jump from the boat and hold her in with the painter while the others, badly frozen, could barely crawl out over the gunwale. He himself and Dunbar in the stern-sheets found themselves so cramped from sitting at the tiller that they could not even crawl and had to be lifted by Starr from the boat. To warm up his men, Chipp had served out immediately two ounces of brandy each, but Dunbar was so far gone that he promptly threw his up and fainted. The second day, under way again, he had

T

kept westward for thirty miles before sighting us in the late afternoon, and there he was, with his crew badly knocked out, in the open water on the edge of the pack surrounding us completely.

To get under way next morning, there was nothing for it but to move our two boats over the ice to where Chipp's was, and with no sledges, we faced that portage over bad ice with deep trepidation. Five men, headed by Nindemann, went ahead with our solitary pickaxe and some carpenter's chisels to level a road. We carried all our clothes and knapsacks on our backs, but De Long dared not take the pemmican-cans from the boats, for so scanty was our food supply getting that the chance of any man's stumbling and losing a can of pemmican down a crack in the ice was a major tragedy not to be risked. So food and all, the boats had to be skidded on their keels over the ice, leaving long strips of oak peeled off the keels by the sharp floe edges as we dragged along. As carefully as we could, all hands at a time on one boat, we lightered them along that half-mile, and when after seven fearful hours of labour we got them into the water, it was with unmitigated joy we saw they still floated.

We made a hasty meal of cold pemmican, and all hands embarked. De Long, last off to board his cutter, was bracing himself on the floe edge to climb aboard, when the ice gave way beneath him, and he went overboard, disappearing completely beneath the surface. Fortunately, Erichsen in the cockpit got a grip on him while he was still totally submerged (for he might not ever have risen except under the widespread pack) and hauled him, completely soaked to the skin, in over the stern. Without delay, except to wipe the water from his eyes, the captain signalled us to make sail.

We fought again fog, ice, and shoals for six hours more to cover the last ten miles to Kotelnoi, and when night finally caught us, all we knew was that we were on a sandbank, where we gladly pitched camp, in total ignorance of whether we had made the island itself or an off-shore bank and caring less so long as we could stop. We found some driftwood on the bank, made a fire, and soon, most of all the captain still in his soaked clothes, we were trying to warm ourselves around it. So ended September 4.

The next two days we tried to struggle west along the south coast of Kotelnoi, largest island of the archipelago, but a blinding snow-storm and ice closing in held us to our sand spit. Going inshore, some of the men found the long-deserted huts of the fossil ivory-hunters, and even a few elephant tusks, but not a trace of game, and our supply of pemmican kept on shrinking.

Signs of physical breakdown were becoming plain enough in our company. Our rations were slender and unsatisfying. Long hours on end of sitting cramped and soaked in wet clothes and icy water, often unable in the overcrowded boats even to stretch a leg to relieve it, no chance in the boats to stretch out and sleep at all, and the mental strain of working those small boats in tumbling seas and through tossing ice, were beginning to tell. On the pack at least, each night we could camp and stretch ourselves in our bags to rest after each day of toil; now except when bailing, we were compelled to endure the cold motionless.

Captain De Long's feet were giving way. Swollen with cold and with toes broken out with chilblains, he could barely move about, and then only in great pain. Dunbar looked older than ever, fainted frequently, and the doctor said his heart showed a weakness that might carry him off under any strain. De Long admonished the ice-pilot to give up all work and take things easy, but even merely sitting up in the boat was a strain which could not be avoided. To keep him braced up, the doctor gave him a flask of brandy with orders to use it regularly. Danenhower's eyes continued the same—poor, but with one eye at least partly useable when the sun did not shine, which fortunately for Dan in all that fog and snow was most of the time. But Dan continued to pester the doctor to put him off the sick-list, driving Ambler nearly wild that he should be nagged to consider such an unethical request. Others, too, began to complain—Erichsen of his feet, Cole of a general dullness in his head. As for all the rest of us, gaunt and underfed, with seamed and cracked faces, untrimmed whiskers, haggard eyes, shivering bodies, and raw and bleeding hands and feet, against a really well man we would have stood out as objects of horror, but there being none such amongst us, our appearance excited no special comment among ourselves.

Our third day on Kotelnoi, we managed to work a few miles to the westward along the coast, rowing and dragging our boats along the sand, making perhaps thirteen miles inshore of the ice-pack which we could not penetrate to the sea beyond. But on September 7, before an early morning north-east breeze with the temperature well below freezing, the pack opened up and we sailed away through drifting ice streaming before the wind, for Stolbovoi Island, sixty-five miles south-west. By noon, I concluded that somehow I must have stove in the bottom of my boat, for we were making water faster than it could be bailed and the boat started to sink. Signalling the others, I hastily ran alongside a near-by floe, where my crew had a lively time getting the whale-boat up on the ice before she went from under us. Capsizing her to learn the damage, I was much relieved to find we had only knocked the plug up out of the drain-hole. We found the plug beneath the overturned boat, tried it again in the hole, and found it projected through an inch. Evidently bumping on some ice beneath the bilge had knocked it out, so I sawed off the projection to prevent a recurrence, righted and floated my boat again. Meanwhile being on the ice, we all had dinner and shoved off again.

With a fair breeze, we stood south-west for Stolbovoi Island, fifty miles off now. The breeze freshened and we made good progress, too good indeed for Chip and the second cutter, as both De Long and I had to double reef our sails to avoid completely losing Chipp astern again. The sea increased somewhat, the boats rolled badly, and we had to bail continuously, but as we were getting along toward the Lena, that didn't worry us, nor did the fact that being poor sailors, Collins, Newcomb, and Ah Sam became deathly seasick again.

We kept on through the night, delayed a bit from midnight until dawn by streaming ice we couldn't see and cold, wet, and wretched as usual. Several times during the night we were nearly smashed by being hurled by surf against unseen floes. Once, under oars, I had to tow the captain's boat clear of a lee shore of ice from which he couldn't claw off, to save him from destruction. But after daybreak, we could see better our dangers and avoid them in time, so that we stood on all day till four in the afternoon, when having been under way thirty-three hours in extreme danger and discomfort, the captain signalled to haul out and camp on a solitary floe near-by. Long before this, we should have hit Stolbovoi, but a shift in the wind had apparently carried us by it to the north.

After a cold night on this floe, at four in the morning on September 9 we were again under way through rain and snow. By afternoon we were picking a path through an immense field of drifting floes which luckily we penetrated and got through to the south-west, when sighting a low island to the westward, evidently Semenovski, the last island of the New Siberian Archipelago between us and the Lena, we abandoned all idea of searching to the southward for Stolbovoi which we had never sighted, and headed west instead for Semenovski. As luck would have it, the wind of which we had too much the day before to suit some members of our party, now died away completely and out went our oars. Through a calm sea we rowed the lumbering boats for six hours, warming up the oarsmen at any rate, though horribly chafing their frozen hands. Then, a fog setting in at 10 p.m., and it being impossible to see the other boats, the captain sang out through the night to haul out on the ice, where by candle-light we ate our pemmican.

Next day, September 10, still rowing through the fog, we made Semenovski by noon, and after a passage of one hundred and ten miles from Kotelnoi, we beached our boats and camped for a much-needed rest. We were all of us stiff, frozen, and sore, but Dunbar especially was quite feeble and looked indeed to be on his last legs.

Semenovski, a tiny island, was to be our last stop before crossing ninety miles of open ocean to Cape Barkin on the Lena Delta, with little chance on that leg of meeting any floes large enough to haul out on for shelter in case the sea kicked up. So while a few men went out with rifles to look for game, we turned to in a final effort to make our boats more seaworthy for this last ocean leg, our experiences so far in rough water strongly indicating the need for improvement. On my whale-boat, I took the canvas boat cover, and by nailing it firmly to both bows and securing it tightly around the mast, I decked over my bow, forming a sort of canvas forecastle. The rest of the boat cover, from the mast aft to the stern, I split in half lengthwise, giving me two long strips of canvas which I nailed fore and aft to the sides. Then making a set of small stanchions which were lashed to the gunwale on each side as supports, I had both starboard and port a flexible canvas weather-cloth eighteen inches high which the men on the windward side could hold up with their backs against the fixed stanchions, in effect raising our rail eighteen inches above the gunwale on either or both sides, but allowing us to drop the weather-screens instantly should it become necessary to get out the oars. Cole and Bartlett did this work on my whale-boat. Nindemann on the first cutter, and Sweetman on the second cutter, fitted them out in a generally similar manner.

While this was going on, our hunters, accompanied by a dozen others as beaters, spread across the narrow island and started to sweep it from north to south. They soon started up a doe and its fawn, which fled in fright, but before long a rifle-shot knocked down the doe which, quickly tossed over a small cliff on to the beach, was brought in a boat to our camp. Needless to say, all else was suspended, driftwood gathered, and at four o'clock, though it was long before our supper-hour, we turned to on a pound of venison steak apiece, which I have little doubt surprised our astonished stomachs, as, accompanied by hot tea, it went down our throats instead of the usual pemmican. That held us until 8 p.m., when we had our regular supper (slightly delayed), consisting of somewhat more than a pound each this time of roast deer, which cleaned up the deer completely except for her bones. Out of these we intended to make soup next day, all except one meaty bone which went to the overjoyed Snoozer. And with that, we felt well fed for the first time since Görtz provided us with bear-steaks a month and a half before off Bennett Island, bear-steaks so far removed from us now in point of time and suffering between, that it seemed almost in a previous incarnation we must have enjoyed that bear!

During our second supper, it blew up half a gale and started to snow, so the captain announced that since the next day was Sunday, instead of getting under way, we would rest on Semenovski, finish our boat-work, and if we could, get that fawn, shoving off Monday for the Lena. I thought this suited all hands, but apparently it didn't, for I heard Collins grumbling to Bartlett:

"Losing over a day for the sake of a feed of meat!"

I looked at the sullen Collins curiously. Whatever the captain did or didn't do was wrong with him. Yet he had downed his "feed of meat" as voraciously as anybody, but perhaps since he expected to taste it again when he heaved it up after we got into the tossing boats, another feed didn't mean as much to him as to a sailor.

Sunday, as on every Sunday without exception which we had passed whether on the pack or in the boats since the *Jeannette* went down, after mustering the crew and reading them the Articles of War, De Long held Divine Service in his tent, attended as usual only by Chipp, Ambler, Dunbar, Danenhower, and myself. Solemnly we listened, seamen about to embark in frail shells for a long and dangerous voyage across the open Arctic Sea, as De Long reverently read the service, and never were we more sincere in our lives than when at the end our rough voices, mingling with the freezing gale howling outside, rose in the final fervent plea:

"Oh, hear us when we cry to Thee
For those in peril on the sea!"

CHAPTER XXXI

AT 7.30 on the morning of Monday, September 12, 1881, ninety-two days since the loss of the *Jeannette,* we shoved off from Semenovski Island for Cape Barkin on the Lena Delta, ninety-six miles away to the south-west of us. The *Jeannette's* company was disposed as follows:

First Cutter	Second Cutter	Whale-boat
Lieut.-Comdr. De Long	Lieut. Chipp	Chief Engineer Melville
Surgeon Ambler	Mr. Dunbar	Lieut. Danenhower
Mr. Collins	Sweetman	Mr. Newcomb
Lee	Warren	Cole
Nindemann	Johnson	Bartlett
Noros	Starr	Leach
Erichsen	Kuehne	Wilson
Kaack	Sharvell	Manson
Görtz	(8)	Lauterbach
Dressler		Aneguin
Boyd		Tong Sing
Iversen		(11)
Alexey		
Ah Sam		
(14)		

There was a fresh east wind blowing, the temperature was just below freezing, and it appeared that we were in for a wintry passage. The island behind us as we drew off was a mass of white snow standing out from the dull grey sea. White-caps were running everywhere. As we had anticipated, there was little floating ice in sight.

For the first hour, we made good progress, shielded somewhat by

Vasselevski Island to windward, this latter being a small island a little to the south-east of Semenovski. By 9.30 a.m., however, we had cleared Vasselevski and received the full force of the sea, careening to it as the boats sped along with the taut sheets singing and our dipping lug-sails drawing full. For two hours we sailed on thus, the first cutter leading, my whale-boat next, and Chipp in the second cutter following me, all the boats tossing considerably. At noon, we found ourselves running again through a moderately open drifting pack, which since we had expected to encounter no ice south of Semenovski, disturbed us exceedingly. But accepting what fate sent us, we seized the opportunity, hauled in alongside a floe, and disembarked for dinner—cold pemmican and hot tea boiled over alcohol stoves. I now had five days' short rations left in my boat, but this worried me little as Chipp and I, stretching our legs on the ice for a few minutes before re-embarking, discussed our prospects. We were both very hopeful; with the wind holding as it was from the east, we should make the last eighty miles to Cape Barkin and the Lena Delta with only one night at sea spent in our boats, and then good-bye for ever to hardship and to pemmican!

"By the way, brother," I asked, looking into Chipp's wan face, so thin now that the resemblance which he once bore to General Grant had completely vanished (unless perhaps Grant also looked like that during his Richmond campaign after the Wilderness), "have you taken aboard yet that can of pemmican De Long is carrying for you? That's your total food supply from now on, you know."

"No, the skipper's still got it," answered Chipp. "But I'm not bothering; I'll get it from him in the morning if we still need it, which I doubt. Don't worry, old fellow," and jokingly he slapped me on the back. "I never expect to have to eat that damned pemmican again!"

"Well, good luck and mind your sailing then, mate." I shook Chipp's hand warmly. "We'll stay reefed down so you can keep up with us."

With a wave, I left him. Chipp was far and away the best seaman of us all—no need to worry over him. I hastened back to my own boat, filled up all our pots with freshly-fallen snow to be used for drinking water on our voyage, and in a few minutes, we all shoved off and were under way again, De Long's last admonition to both of us being to keep formation astern of him and hold our little squadron together.

The wind continually freshened and soon hauled to the north-east, dead astern of us. This made the sailing more hazardous, for we now were constantly exposed to the danger of jibbing our sails, but for some hours more, we bounced along over rough seas, on the whole grateful for the occasional drifting floes we encountered because they tended to break the waves. I managed to manœuvre safely amongst these floating menaces until four o'clock, when following the first cutter through a narrow passage between two floes, a wave hurled my whale-boat to leeward, staving in our starboard side against sharp ice and I hurriedly had to haul the boat out on the floe to keep from losing her. In fifteen minutes, while the other boats lay to in the lee of that ice, I had her repaired by tacking a box cover over the hole, and overboard again, we filled away to the south-west in regular formation. That (though I soon had cause to regret it) was the last of the ice-pack we ever saw.

In the open Arctic finally, we ran on, the breeze freshening all the time from the north-east and the sea picking up. Before long, Chipp began to drop steadily astern, since both De Long's cutter and my whale-boat were far better sailers than his short cutter. To avoid losing him, we had to reef sail, first taking one reef, and soon a second, after which both the leading boats ran close-reefed while Chipp, with his sail full out, barely managed to hold position in column in my wake.

Jack Cole, who was my coxswain, had been steering since morning. Jack, put in my boat by De Long because he had been such an excellent small boat sailor, was, however, now a severe disappointment to me. For some days he had been dull and apathetic, seeming hardly to know what was going on in the boat. The weather we were facing required prompt and vigorous action at the helm, so with some reluctance I relieved Cole of the tiller, replacing him with Seaman Leach, and detailing Manson and Wilson, good sailors both, to take the tricks following, intending to relieve Leach in four hours.

As the late afternoon faded, the wind whipped up to gale force. Spray came in over our stern, soaking all of us through and through, while the freezing wind chilled us to the bone. Wet and miserable, buoyed only by the thought that before that gale we should certainly make the Lena and safety by morning, we struggled to hold position, my whale-boat pitching and rolling badly, while ahead and astern of us, we could see the other boats heaving even more crazily in the seas, a sight which did little to encourage us. To keep from jibbing, running free as we were, I was forced to station a man with a boat-hook to hold out the sail; and not daring to trust anyone else with the job, I manned the sheet myself, hour after hour clinging to that freezing manila line. Heavy spray began to break over our stern, and we started bailing.

But despite close reefs and frequently dousing sail to deaden headway, we began to run ahead of the first cutter, which heavily laden as that square-sterned boat was with the records of the expedition, its stores, and more men than any other, proved a slower sailer than my double-ended whale-boat. Vainly I tried to hold astern of the first cutter, but each time I doused sail, the racing seas combing over our stern came heavily aboard, forcing us to bail vigorously to avoid waterlogging. By seven o'clock in the fading twilight with the wind blowing a lively gale and the sea, already bad, rapidly getting worse, I found my little boat in spite of all my efforts a thousand yards out on the weather bow of the first cutter and steadily gaining on her.

At this moment, Manson sang out:

"Chief, Ay tank dat cutter ban making signals to us!"

Looking astern across the waves, I saw De Long waving to me, apparently to come within hail. We were already close-reefed and there was no way to achieve this except by dousing sail again and drifting while he caught up, so I stationed several men to gather in the foot of the sail (she was rigged with a single mast and a dipping lug-sail) while the yard came down. We partly doused sail and slackened speed, but as we did so, a sea caught us and boarded our stern, flooding us all to our hips in icy water. The men holding the foot of the sail, startled at the fear of swamping, promptly let go and we had for a moment the chaos of a flooded boat erratically heaving in the seas, a flapping sail threatening to take the stick out of her, and every man bailing wildly with any utensil at hand that would hold water.

We finally got her cleared and the sail hoisted to get some headway and hold us before the seas, when we tried again. Once more we were flooded, but while some bailed, the men at the sail, strongly admonished to hang on no matter what happened, clung to it this time, and we managed to drift almost within hail of De Long. I could see him shouting to me from his cockpit, but whatever he said (I being to windward of him) was lost in the roar of the gale. Just then another sea rolled up and combed over both boats, nearly filling mine. Instantly I jumped to my feet and bellowed down the wind to him:

"I can't hold back, captain! It's either run or swamp!"

Perhaps he heard me. I have a strong voice and I had the wind behind me to carry my words along. At any rate, he could see the situation, and with another shout, also smothered in the wind, he waved energetically, motioning me on.

I needed no more. Promptly loosing sail, we filled away, still bailing our waterlogged craft, while I waved to him in acknowledgment. De Long turned in his tossing cockpit, and I then saw him motioning violently to Chipp, half a mile astern, possibly wanting him to come close enough aboard to toss over that forty-five-pound can of pemmican, the only food Chipp's men would have, before the gale and the darkness separated them.

But I had my hands full in my own boat and paid little attention. We hoisted sail, shook out one reef, and shot ahead down the wind before another sea could catch us and finish the job of swamping us completely. Having gathered sufficient headway to manœuvre a bit, I hauled the boat a few points closer to the wind, so that instead of heading south-west dead before it with the consequent grave danger of jibbing the sail and broaching to before the oncoming seas, we now ran with the wind on our port quarter, heading roughly south and driving hard amidst heavy seas. We had our canvas weather-cloths up on both sides, with the freezing men in the boat, their backs against the cloths to hold them in position, themselves all standing poised with pans ready to bail in the brief intervals when not actually bailing. And indeed, had it not been for those canvas weather-cloths, so carefully fitted the day before at Semenovski, we should long since have foundered. As it was, we huddled behind them, all save Leach at the tiller, with huge seas rolling past our raised sides and sweeping heavily along the billowing canvas screens, over which even so, spray and some solid water from every crest dashed into the boat.

The sea was now running mountains. Our little boat was tossing wildly, rising dizzily to every crest as it swept up, then plunging madly down into the trough as the wave rolled by. The wind roared on, icy spray, cutting like a knife, drove into the boat, our sheets and halliards sang in the gale, while the mast in its step creaked dismally and our yard whipped so violently in spite of a double-reefed sail, that with each gust as we rose on a crest and the wind caught us squarely, I began to fear that both mast and sail would go flying from us down the wind like a suddenly released gull.

"Look!" shouted Leach at the tiller. "The skipper's signalling us again, chief!"

"Never mind him!" I growled. "Watch your steering, Leach. We're on our own now. Nobody sees any more signals for us!"

But nevertheless I looked aft myself through the twilight. We were fast outdistancing the first cutter where half a mile to windward already, I could see De Long gesticulating in the stern. But Leach was mistaken; De Long seemed to be waving to Chipp a thousand yards to windward of himself, and now a mile astern of us. As I stared, shielding my eyes as best I could from the sharp spray, I saw the second cutter rise against the sky on the crest of a breaking wave, then sink into the trough. Again she rose, when an immense sea swept over her and she broached, lying helplessly broadside to the gale! Instantly her sail jibbed and the yard swung over, binding yard and sail against the mast. A man sprang up, sharply outlined against the horizon, struggling frenziedly to clear the jammed sail from the mast, then the heeling boat plunged broadside from my sight into the trough. The second cutter had evidently swamped!

Suddenly sick, I watched that spot as wave after wave rolled by, but nothing rose again, and only flying foam and breaking seas met my gaze. Broken-hearted, I stared across that mile of raging sea at the scene of that swift tragedy. There was nothing we could do. No boat could ever beat a mile dead to windward against such waves; long before we could even get our boat into the wind on the first tack, the icy waters and the tumbling seas had ended the agony of the men in the second cutter.

I sank back in the stern-sheets, sobbing for the shipmates I had lost. Quiet, taciturn Chipp, who by sheer will-power had conquered sickness to lead his men across the pack; grizzled old Dunbar, who had broken his health scouting paths for us over the ice; huge Starr, whose herculean back had many a time lifted my jammed boat over the hummocks in the pack; little Sharvell, whose comical seriousness had often lightened our months of tedious drifting; Sweetman, Warren, Johnson, Kuehne, good seamen every one—their struggles were for ever ended. All their agonising labours to escape from death in the pack had brought them only death in the foaming waves. Now their voyage over, they were slowly sinking through the cold depths to unmarked graves in the desolate Arctic Sea!

Chipp, the best seaman of us all, had swamped in the gale, and we were soon like to follow, for what chance was there for us where his skill was insufficient? But at least we in the whale-boat were still afloat. What of the first cutter? Fearfully I looked aft again for her.

Still half a mile off our starboard quarter, there she was, plunging furiously along before the tumultuous waves, holding grimly to her original course, south-west, dead before the gale for Cape Barkin. I shook my head sadly. De Long should haul closer to the wind. As he was heading, it would be a miracle if the square-sterned first cutter lasted ten minutes, for one bad jib would dismast his boat and broach her also, needing only another wave to send her tumbling on her side to follow in Chipp's wake. But it made little difference. Regardless of course now, every sea combing past seemed ready to swamp our two insignificant cockle-shells. Chipp was already gone. Any minute now I confidently expected to feel that strangling water in my throat, after a few feeble struggles to be overwhelmed by the breaking seas and, still gasping for air, to sink numbly through the frigid water to join Chipp on the bottom.

I saw no more of the first cutter. On different courses, she soon faded from our view into the night, and we were left alone, eleven freezing men huddling in a tiny whale-boat in a world of roaring winds, of mountainous seas, and of utter blackness.

We drove along before the storm, the wind on our port quarter, the rushing seas pressing madly against our port side, with the canvas weather-cloth there billowing and sagging inboard as the crests swept along the canvas and poured in over the top. Each time we slid sickeningly down into a trough, all hands bailed for dear life, fighting to get the flood sloshing round beneath the thwarts overboard before we rose to the next crest and more water pouring aboard from the succeeding wave swamped us completely. Every pot and pan we had in the boat was pressed into service, and except for Leach, who, braced in the stern-sheets, clung to the tiller and steered, and except for me manning sheet and sail myself to keep us from capsizing, the others in the boat alternately pressed their shoulders back against the weather-screen to hold it up while rising to the crests, and leaned forward, bailing furiously while sinking into the troughs.

Numbed fingers clung to ice-coated pans, icy water sloshed over our heads and down our necks, our frozen feet had long since lost all sensation, and the careening boat beneath us pitched, rolled, and heaved so dizzily that only by clinging continuously to the thwarts did we manage to escape being tossed bodily overboard.

Leach in the stern, the one man besides myself whose task required him to stand motionless in one spot, did a miraculous job of steering. Had he but once allowed her to swing off the course we should have broached immediately and capsized. But clinging to the tiller in the darkness, more by feel than by such vague sight as the foaming crests sweeping by gave him of the direction of the sea, he kept the wind on the port quarter, standing himself wholly unprotected, a fair target for each smashing wave breaking over the stern to drench him completely.

Straining my eyes through the blackness as I clung to the sheet, I watched the seas coming over and the men bailing. It was obvious that we were fighting a losing battle. Even though Leach at the helm and I at the sheet managed to keep her from broaching or jibbing or both at once, and thus instantly ending the struggle, it was evident that those bailing could not indefinitely keep ahead of the water coming in over our stern, and sooner or later we should fill and founder. Oh, for some drifting floes! If only we might run into another field of ice which would at least deaden the seas if not allow us to haul out on a floe! The ice-pack, which many times in the long months past I had cursed vehemently, I now earnestly prayed for as our only sure salvation. But we had long since dropped astern the last floe, and now we must battle it out with the turbulent sea. There was only one thing more I could do before my men, barely able now to keep us afloat by bailing, between numbness and exhaustion found themselves unable to keep the water going overboard as fast as the tempestuous seas poured it in. Regardless of the hazard involved, I must come about, head into the crests, heave to, and ride out the gale bows on, held that way by a sea-anchor, while our canvas-decked forecastle, instead of our open stern, took the brunt of the oncoming waves.

But to heave to meant inviting quick destruction, for the manœuvre involved turning broadside to the waves for an instant as we swung our bow about to head into the wind. To make matters worse, while in that critical position we must use our oars to swing the boat, and that meant dropping our weather-cloths down to the gunwales at the very instant when most of all we needed every inch of freeboard possible to avoid taking over a solid beam sea and foundering out of hand.

Just abaft me, huddled against the weather-cloth, was Danenhower. I leaned over in the darkness and shouted into his ear:

"Dan, it's blowing like blazes! We'd better heave to!"

Dan turned his face, with his bandage dripping water, toward me and nodded:

"Yes, Melville. You should have done it long ago!" he yelled back.

That startled me, for if he thought so, Dan should certainly have made the suggestion himself. Why had he chosen instead to keep his mouth shut? But it was no time to argue why or wherefore. We must heave to.

"Get hold of Jack Cole!" I roared. "We'll try it now!"

Soon in the stern-sheets, Cole, Danenhower, and I were debating what we should use for the sea-anchor, or drag, to which the boat must ride. Dan advised making the drag of three of our oars with our sail lashed between them, but this I refused to do, for if we lost the sea-anchor, our oars and sails would both be gone and then we would be helpless indeed. Canvassing what little else we had in the boat to stretch out the sides of a drag, I hit on three brass-tipped tent-poles, and these I ordered to be used, with a section of tent cloth as the drag itself.

Working in the half-swamped boat, Cole and Manson together made the drag, lashing the ends of the three tent-poles into a six-foot triangle, and then lacing inside it the piece of tent. Meantime Danenhower unrove a small block and fall to get a line, and from that manila-line made up a short three-legged bridle, one leg of which went to each corner of the triangular drag, while all the rest of the line was to serve as our anchor cable. The result of all this, when we were through (which working under bad conditions in the darkness took two hours) was that we had what looked like a large triangular kite at the end of a long manila-line, the main difference being that our kite or sea-anchor we were going to fly in water instead of in air, and as it dragged vertically through the water, it was to hold us head to the oncoming seas that we might ride them bows-on.

On one matter at the end of all this, Cole and I differed. The drag had to be heavy enough to sink beneath the surface, for if it floated, it would be ineffective. But if it were too heavy, it would sink too far and instead of streaming out ahead of us, would hang

U

vertically beneath our bows, keeping them from rising to the seas, and helping to swamp us. Cole stoutly maintained that the completed drag was not heavy enough to sink. While somewhat inclined to agree, I refused to add more weight, for to a buoyant drag I could always add more ballast, but it was doubtful in that storm that I could ever get my hands again on a sunken drag to remove excess weight, until the boat having swamped, I caught up with the submerged drag on my own way to the bottom. However, to appease Jack, I got a copper fire-pot ready to slide out on the line if it should be necessary to add more weight, and we were ready to proceed.

At this point, Lieutenant Danenhower, who had been busy making the bridle, spoke up again.

"Melville!" he sang out in my ear above the howling wind. "Will you let me heave her to?"

I hesitated. Dan was over half blinded. But, I thought to myself, in this darkness that was no great handicap; he could probably see as much as anyone. And certainly as a deck officer, he should know more about handling a boat in a seaway than anyone else in her, including me, an engineer. Our lives depended on that manœuvre; I should be derelict in my duty not to use the best talent available.

"Sure, Dan," I replied, "go ahead and give the orders. We'll all do what you say."

So to poor Danenhower, for nearly two years now nothing but a helpless burden on the rest of us, had at last come opportunity to serve. The crisp way in which he rattled out the orders, stationing the men, infused new strength into the exhausted crew. To stocky Bartlett went the most important task; he was posted in the bow with the drag, and the anchor-line was carefully coiled down at his feet, ready for running freely out when he heaved the drag over at the word. Bosun Cole stood by the halliards to douse sail, with Aneguin and Tong Sing to gather in the canvas as it came down. I tended the sheet, Leach steered, Manson stood by an oar poised high on the port side, and Wilson an oar similarly ready to starboard. This left only Lauterbach to tend to the bailing, for Newcomb, as always, was next to worthless, even though he now had the excuse of extreme sea-sickness to mask his piddling efforts at real labour.

With a last all-hands dash to bail down as dry as possible, the men took their stations. In silence and in darkness we waited, holding up the weather-cloths while several tremendous waves rolled by and we rocked wildly to them. Judging then that we should have a momentarily quieter sea, Danenhower bellowed:

"Out oars! Starboard the helm! Down sail!"

Down dropped the weather-cloths, out shot the two oars, hard a-starboard went the tiller, and down came the flapping sail and yard as we swung off to port, trying quickly to come about and head up into the sea. Wilson to starboard gave way valiantly on his oar while Manson to port backed heavily on his, endeavouring to get her spun about before the next wave caught us broadside, when with the boat hardly more than a quarter way round—Crash! came a wave tumbling in over our low gunwale and flooding us to the thwarts!

Heaving to was forgotten. With might and main all hands dropped their tasks to start bailing again except Leach and the oarsmen, who struggled desperately to straighten her out again on her old course before the waterlogged whale-boat broached to the next wave and capsized. By the grace of God we succeeded in that, and hearts in our mouths, bailed madly while with the oars alone we kept away before the sea till the boat was sufficiently dry to risk coming about once more.

After a hurried consultation with Dan, I stationed the men as before, and we stood tensely by. Again we pitched crazily to a succession of roaring crests, and as we slid dizzily down the trough of the third one, Dan once more sang out the orders.

This time, with oars thrashing the sea in desperation, we swung more quickly, and before the next wave struck, we had the boat spun about, head on to its breaking crest as our bow suddenly lifted to a tremendous wave.

"Let go the drag!" bawled Dan, and Bartlett in the bow tossed over the canvas triangle. At that instant, the bow dived sickeningly into the next trough, and Bartlett, off balance, pitched headlong forward toward the sea!

For an instant I thought he was gone completely with little chance of our getting to him before the sea swept him away, but fortunately our slacked halliards were streaming out over the bow, and as Bartlett shot overboard he got his right hand on the halliards, stopping his plunge. The next second, as he hung there in air, the boat rose again to the crest of a huge wave, the mast whipped back, and Bartlett came flying inboard on the halliards against the mast to which he clung for dear life as he slowly slid down to the thwarts.

Meanwhile other things were rapidly happening. Still manning the oars, Manson and Wilson were holding us head to the seas while the drag-line ran out and I watched it anxiously till it brought up at the bitter end. In disappointment, I saw that the drag was too light, coming immediately to the surface and drifting down to leeward, holding us not at all. We yawed badly, shipping water over both sides in spite of all our two oars and the rudder could do to hold us bows on, and, as expected, Jack Cole immediately piped up with:

"Shure, Mr. Melville, I tould yez so!"

"Right, Jack; you did!" I shouted. "And now we'll fix it. Bartlett! Send down that fire-pot!" Bartlett, again on his feet, seized the copper fire-pot and sent it sliding on a lashing out over the bow and down the anchor-line to the drag, which it promptly sank. A heavy strain came immediately on the drag-line as the sea anchor gripped the water and in a few seconds we were riding head to the seas with our oars in, our weather-cloths once more raised on our bulwarks, our sail furled, and the helmsman, as before, continuously steering to keep us from yawing on the drag-line.

My thumping heart quieted somewhat. Dan had done a fine job. We had come about without capsizing, and for the moment we were safe. All we had to do now was to bail continuously to keep afloat, but so long as our drag-line held, we were secure. But as much vigilance as ever was necessary, for if that line parted, and we went adrift, unless we were immediately ready with the oars, our first yaw would probably also be our last one.

By now it was ten o'clock and pitch dark. All through the rest of that terrible night we tossed violently at the end of our sea-anchor line, bailing, always bailing. My hands were swollen by cold, blistered from hanging on to the sheet while we sailed, and cracked and split by freezing salt water, while my feet were both badly frozen. Leach, who for the first time I now dared to relieve from his station at the tiller, was as badly off as I, and the rest of the crew, not much better. And now thirst was added to our sufferings, for we had not a drop of fresh water, every bit of snow that originally we had in the boat in our pots and kettles having long since been thoroughly soused with sea water and spoiled.

The gale shrieked on, the waves rolled by, the cold spray dashed in and froze on us, and in the darkness wearily and endlessly we bailed the boat till at last came dawn to put the final touch to our misery by adding to each man's sense of his own suffering the sight of his shipmates' wretched state.

Cold, thirsty, and hungry after twelve excruciating hours of bailing, my exhausted crew looked expectantly at me for their rations. In the boat, there was nothing left in the way of food or drink but a little pemmican.

Sadly I recalled Chipp's last remark about his being through with pemmican, only a jest when made, tragic now. Poor Chipp! Pemmican, in truth, meant nothing to him any more. If only I had him and his second cutter's crew following astern of me again, how gladly would I divide my meagre supply of pemmican with them! Mournfully I looked at our tiny stock. At the previous miserly rate of issue, it should last us five days, but no longer did I dare to hope that Cape Barkin and rescue were just over the horizon. Heaven alone knew where, drifting at the mercy of wind and wave, we would be when the gale blew out. I must stretch our food to the utmost. So in spite of grumbling from my ravenous crew, I cut our already short ration squarely in half and issued for our breakfast so small a piece of pemmican that not a man, after swallowing his ration at a gulp, but growled for more.

At the end of that sea-anchor line, our whale-boat weaved, twisted, and leaped erratically about amongst the foaming waves, more terrifying now that we could see the thundering crests sweeping down and breaking over us, than even in the darkness. Monotonously we bailed; Aneguin, our Indian hunter, and Charley Tong Sing, our Chinese steward, strangely enough proving far more dexterous with the bailing-pots than any sailor in the boat. Almost helpless myself, with numbed hands and frozen feet from long hours of hanging on to the sheet, I watched forward while the endless task went on, keeping a weary eye on that thin manila line going over our bow to our sea-anchor, that thread to which our boat rode head to the gale, our life-line indeed.

And then mixed with the screaming of the winds, I caught a burst of laughter. Laughter? I could hardly believe my ears. What in God's name could anyone see in our situation that was funny? Painfully I twisted my ice-sheathed body round on the thwart to see, and then I groaned. On the midship thwart sat Jack Cole no longer bailing, a maniacal gleam in his eyes, rocking with childish glee in the spray of each wave as it broke over the weather-cloth and poured in on him, laughing, laughing horribly. Jack Cole, our bosun, like a little child was splashing playfully in the water!

Like a flash there came over me an understanding of Cole's apathy and inexplicable stupidity during the last few days. His mind had evidently been going then; now after the horrors of the night, it was completely gone!

That terrible laughter must not continue. With my whole crew near the breaking point, a little more of that insane shrieking in the storm and I would have only a boat-load of lunatics to depend on, if indeed I did not soon become one myself. I dragged aft a bit, took Cole gently by the hand.

"Come on, Jack," I said kindly, "let's go forward where we can see the waves better and I can help you watch them."

Jack stopped laughingly, looked blankly at me, but offered no resistance and together we dragged our frozen legs over the thwarts to the bow. There, smiling happily, Jack started laughing again as a terrific sea broke dead on our stem nearly drowning me in foaming water.

In the midst of that insane cackle, I grabbed Cole suddenly by both shoulders, pushing him heavily down beneath the thwart and shoved him in under the canvas-covered forecastle forward of our mast, where at least the canvas would muffle that ghastly laugh.

"Take a nap now, old fellow," I said soothingly as possible. "You're tired!"

"Shure an' I am that," mumbled Cole beneath the canvas. "But ye'll not be fergettin' to call me agin soon so's we kin watch thim waves?"

"No, Jack, I won't," I promised. "Just as soon as you're rested."

Without another word, Jack, half-covered with the icy water sloshing about in the bilges, went dead to the world, and I, turning wanly from the gale, sagged back against the mast, staring at the haggard men in the boat abaft me bailing, bailing, while the never-ending waves foamed up and broke sickeningly over our bow.

CHAPTER XXXII

AFTER four harrowing days in the whale-boat, the first two in the gale, the second two fighting our way through off-shore shoals in the open ocean, we finally sighted land. Hungry, thirsty, frozen, we gazed as hopefully across the sea at two low headlands barely showing above the horizon as though they were the very gates of Heaven. Under oars, for the wind at last had died away, we propelled our boat toward them, and soon found ourselves between two low hills apparently forming the mouth of a wide and muddy river running swiftly out into the sea. All hands leaned over the gunwales, and finding the water sweet, we drank greedily, regardless of mud, regardless of everything.

Where were we? Neither Danenhower nor I knew, but as that low and barren coast trended north and south, I assumed we were on the eastern side of the Lena Delta, how far south of Cape Barkin I could not tell. But at least we had Siberia at last before us! My orders directed me to land at Cape Barkin, where I should find native huts, but I had had enough of the Arctic Sea for the present and for the future, and with Barkin an unknown distance up an unkown coast, I decided to be satisfied with the land I could see and proceed up the river before me till I located some village there. In spite of Danenhower's objections to my course, we rowed (that is, if the feeble efforts of the half-dead sailors manning the oars could be called rowing) up the broad river, constantly attempting to make a landing on either bank, but always baffled by shoals, which prevented us from getting within a hundred yards of those flat and muddy shores.

Finally, in the late evening after a gruelling day at the oars, we spotted on one bank an abandoned hut, before which was a cove into which we made our way thankfully, and for the first time in five days landed to stretch our legs. I found that I could hardly move mine; most of my men were in like case. Only Danenhower, whose blindness had excused him from bailing, and Newcomb, who had most successfully evaded it, had managed to keep their legs in shape so that they could walk. The rest of us had practically to crawl from the boat.

Thinking to warm up and thaw out our blackened and frostbitten feet, we gathered driftwood, made a fire in the hut, and huddled round it, stretched on the ground, with our numbed feet toward the fire. But instead of helping, agonising pains started to shoot from my paralysed feet as soon as the heat took effect, so stripping my legs for examination, I found they were frozen from the knees down, terribly swollen, covered with cracks, blisters, and sores all run together, and with the skin sloughing off at the slightest touch. Excruciating pain instead of sleep was my portion our first night ashore, and in place of the eagerly-awaited comfort which we had looked forward to in Siberia, most of us writhed in pain, suffering the tortures of the damned.

At dawn, after a slim portion of pemmican washed down with muddy tea, we launched our boat and set bravely out up the river to find a village, only to discover instead that we were in a desolate maze of shoals, swamps, and muddy islands forming the delta, with rivers, sometimes swift and sometimes sluggish, criss-crossing erratically as they flowed over the low delta lands to the sea. Young ice was forming everywhere over river and swamp and through it with boat-hooks and oars we had to smash a way for our bow. Three days of this we had to endure, alternating between slaving at the oars during the day and freezing at night in our camps on the barren mud flats, while both night and day we starved on scanty rations, and I finally began to despair of rescue. Here we were on the Lena Delta, but of the many villages indicated on Petermann's charts, we could find no sign. Never a native did we see, and the few huts we spotted now and then were all abandoned, their owners having already retreated southward before the oncoming winter, which was rapidly robbing us of what little vitality remained in our feeble bodies. Were we never to escape? Were all our sufferings to end only in our deaths in the delta? Had we not already borne enough since those harrowing years on the *Jeannette* to be spared that? First the torture of dragging boats and sledges over the pack, then the horrors of navigating amidst the streaming ice of the New Siberian Archipelago, finally that four-day nightmare of tumbling waves and freezing spray in the open whale-boat battling an Arctic gale—was all this not enough? Yet through all our trials since the loss of the *Jeannette* we had been sustained by the thought that if only we held out till we reached the Lena Delta, there at last our sufferings would end, amid friendly natives we would find food, shelter, and transportation home.

How different now was the reality! The Lena Delta we found a bleak and barren tundra, empty of game, as inhospitable and as desolate as that ice-pack in which for two years we had drifted in the long-lost *Jeannette*. Our dream of a safe haven had exploded in our faces. With food gone, men worn out, and worst of all, the hope which had driven us all to superhuman labour proved a lie, our situation was desperate beyond conception. Bitterly we cursed Petermann and all his works, which had led us astray.

But there was nothing to do save to move on, working always toward the headwaters of the delta as long as we could swing the oars, so for the fourth day in succession, we shoved off from a mud-flat camp, broke our way through new ice, and I pushed my men (whose arms fortunately were a little better off than their legs) upstream toward the delta head.

And then, thank God, in the middle of this day, while deadened arms and stupefied bodies swung wearily over the oars, we suddenly sighted three natives in kyacks shoot out from behind a bend in the swamp!

Like drowning men grasping at straws, we waved to them, shouted to them, and tried to row to them, but before the apparition of a strange boat in their waters, they were shy and afraid, and not till I held up our last tiny strip of pemmican did I entice one, more curious than his comrades, close aboard us to taste the strange meat. Then like the jaws of a trap closing on its victim, we grabbed his kyack before he could dart away!

Badly frightened, the fur-clad native attempted to escape, but we would sooner have released our only hope of salvation than our grip on that poor Yakut who represented now our last slim chance to avoid perishing in the maze of that frozen delta, and we held to him like grim death. Gradually I calmed his fears, gave him the pemmican, endeavoured in pantomime to show him we were friendly, and at last holding to him while we beached our whaleboat, convinced him of our good intentions by giving him a little of the trifling quantity of alcohol we had left for our stove.

The alcohol settled the question. He promptly hailed his two comrades standing warily off in their kyacks, and soon all three of the natives, warming up on pure grain alcohol, were our bosom friends. In exchange for the alcohol, they gave us some fish and a goose out of which mixture we promptly made a stew which we wolfed down ravenously. And then with pencil sketches and gestures, I endeavoured to make plain that I wanted them to guide us to a village, and specifically to Bulun, the largest town shown on my chart, some sixty miles up the Lena River from the head of the delta.

It was remarkable how, understanding not one word of each other's language, we got along. The three Yakuts indicated we could not get to Bulun on account of the ice in the river, that we should all die on the way. However, they made plain that another Yakut village, Jamaveloch, they could take us to, and next day for Jamaveloch we started. But so tortuous was the course and so hard the labour in working our boat through the delta swamps and rivers, that not till a week later did we finally, on September 26, two weeks after the gale, arrive at Jamaveloch. Had it not been for the food provided by our guides as well as for their pilotage, it is inconceivable that we should ever have arrived alive at this village at the south-eastern corner of the delta, seventy miles from Cape Barkin, and the only inhabited village for over a hundred and fifty miles in either direction along the Siberian Coast! Had we gone to Barkin, we should assuredly have perished, for there, the natives told us, Petermann was absolutely wrong—there were no villages, no lighthouses, no inhabitants of any kind there, nothing but a barren coast.

But Jamaveloch itself was not very promising as a haven except for a brief stay. It had but six huts and a few small store-houses, not over fifteen adult inhabitants, and no great surplus of food. Doubtful that its scant supply of fish and geese would long take care of eleven voracious seamen thrown unexpectedly on the resources of so small a community, I decided after one night at Jamaveloch to push on in our whale-boat up the south branch of the Lena to Bulun, a hundred and ten miles away by land but a hundred and fifty miles distant up the winding river. Strenuously in their native Yakut tongue (some of which I had now picked up) the villagers and especially their headman, Nicolai Chagra, objected that ice in the river would block us and leave us to perish along the uninhabited river banks, but I persisted. So accompanied once more by my original native pilots, I loaded my sick crew into the whale-boat, took aboard sixty dried fish (all I could get) for supplies from Nicolai, and we started. In an hour we were back. Nicolai Chagra was right. The *Jeannette* herself could not have ploughed through the ice, alternately freezing and breaking loose in the river, which swept downstream in the current, effectively blocking any progress toward Bulun.

Willing or not, there was no choice but to stay at Jamaveloch. Unable to walk, I crawled from the whale-boat and was hauled on a sledge from the shore to a hut turned over to us by Nicolai; Leach, with the flesh falling from his frozen toes, was hauled up on another sledge; and most of the rest of my crew in the remnants of their tattered clothes, crawled or hobbled after us.

For two and a half weeks we lay in that hut, slowly recuperating from our frost-bites, subsisting mainly on a slim ration of fish given us daily by the Yakut villagers, and thinking up weird schemes of getting away to Bulun. But till the rivers froze solidly enough to sledge over the ice, there was no chance. Even then, the limited facilities of the village could never provide the necessary sledges for eleven men nor the clothes to keep us from freezing in the sub-zero weather which October had brought. But get away soon we must, for all the flesh had sloughed from several of Leach's toes and he needed medical attention badly if he were not soon to die; while Cole, lucid at intervals, required expert care also if his mind were to be saved; and Danenhower's eye, a month now without surgical care, was beginning to relapse. As for the rest of us, our legs were getting better and we could soon drag ourselves about, but the food problem was rapidly getting acute, and I was very much afraid that we should awake some morning to discover that the natives, finding us too much of a drain on their stores, had silently moved on in the night to some other collection of vacant huts of which we knew nothing, leaving us to starve alone lest everyone starve together.

The only solution to this dilemma, since we could not go to Bolun, was to have Bolun send us the necessary dog-teams, sledges, clothes, and food to make the journey. How to get word to Bolun, however, was the difficulty, for none of the natives would go and no man in my party knew the road over the distant mountains to Bulun. I dared send no one without a guide.

The reason given by the natives for refusing to undertake the trip was that it was an impossible season for travelling, an in-between time in which they could safely move neither by boat nor sledge. A few weeks before, in early September, it would have been possible to go by boat, but now new ice forming everywhere prevented. A few weeks later, it would be possible to travel by sledge cross-country over snow and ice, but just now that also could not be attempted for the ice on the many rivers to be crossed was continuously breaking in the current and was nowhere yet thick enough to bear the weight of a sledge without grave danger of crashing through into the river and losing sledge and dogs at least, if not drivers also. To all our entreaties, Nicolai Chagra merely shrugged his shoulders—early September, yes; late October, yes; but now, a most decided no!

Providentially the matter was settled for us about the middle of October by the chance visit to the village of a Russian exile, Kusmah by name, who lived near-by and who on the promise of the whale-boat immediately and five hundred roubles later (when I could get funds from America) undertook to make the dangerous journey and started off with his dog sledge over the frozen tundra to Bulun, expecting to return in five days.

Vaguely, while he was gone, we speculated on how long it would take us to sledge the fifteen hundred miles from Bulun -via Yakutsk to Irkutsk on Lake Baikal, and then via post road get to Moscow and so home. And while we speculated over that, we also speculated earnestly over the fate of De Long and the first cuttter. There was no doubt that his boat had followed Chipp's, but over the question of how long the first cutter had lasted in the gale and whether she had come to her doom finally by capsizing or by swamping, there was many a hot discussion, as my seamen argued vehemently over the relative probabilities of a square-sterned boat like the heavily-built first cutter broaching before she flooded, or vice versa. The consensus of opinion was that she had swamped, for De Long had in his boat not only three more men than we, but also Snoozer, the last dog, all the navigating equipment, four rifles, the complete records of the expedition in ten cases, and one small sledge which De Long had kept to drag the records on. With so much ballast in his boat, that his men could have bailed fast enough to avoid foundering seemed incredible to most of us after our own experiences with the much lighter double-ended whale-boat, but the broaching theorists would never agree to it. Chipp, whom all hands freely admitted was the best sailor, had broached and capsized. How then could De Long have avoided it? And since, crowded in our little hut with nothing else to do, there was no outlet for men too feeble to get about save in talk, the argument went on endlessly, and of course with no chance of an agreement ever being reached.

Five days went by and Kusmah, our messenger to Bulun, had not returned. Ten days elapsed and we became alarmed for Kusmah. Had he perished in the ice? To add to our worries, Nicolai Chagra cut our food supply from four fish a day to three, with occasionally a putrid and decaying goose supplied in lieu of the fish.

I was seriously debating sending Bartlett, the strongest member of our party, on to Bolun in the forlorn hope of getting us assistance, when on the night of October 29, after thirteen days' absence on his hazardous journey, Kusmah at last returned, bringing on his sledge some supplies, about forty pounds of bread mainly, and no clothes for us, but instead a letter in Russian from the Cossack commandant at Bulun stating that next day he could start for us from there with a reindeer caravan and clothes enough to bring us all safely over the mountains to Bulun.

This news heartened us considerably, and in broken Russian I profusely thanked Kusmah. Meanwhile, my men, not waiting to thank anybody, were revelling in the bread of which we had seen none for nearly five months, breaking the loaves in huge chunks into which they sank their teeth hungrily. All smiles at my expressions of approbation, and happy at the way everyone seemed to appreciate what food he had brought us, Kusmah bowed, then pulled from inside his fur jacket a dirty scrap of paper which he tendered me. On it was a pencilled message. Pausing casually between two mouthfuls of bread, I glanced at it, noted in surprise that it was in English, and then as I read the first words, I stiffened as suddenly as if I had been shot.

"Arctic steamer *Jeannette* lost on the 11th June; landed on Siberia 25th September or thereabouts; want assistance to go for the CAPTAIN and DOCTOR and nine (9) other men.

WILLIAM F. C. NINDEMANN,
LOUIS P. NOROS,
Seamen U. S. N.

Reply in haste; want food and clothing.

For a moment my heart stopped beating as I read, then I called out huskily:

"Men! De Long and the first cutter landed safely! They're alive!"

All over the hut broken loaves of bread thudded to the floor as open-mouthed in astonishment at this startling declaration, my shipmates stared at me, then clustered round to read the note, while I turned abruptly to Kusmah, asked in my best Russian:

"That note, Kusmah! Where did you get it?"

With some difficulty, Kusmah explained to me his trip. To get to Bulun, he had to go fifty miles due west across country over the mountains to Ku Mark Surk on the Lena River (where he was delayed a week waiting for the main stream to freeze over so he could cross) and then sixty miles due south along the west bank of the Lena to Bulun. On his way back to us from Bulun, coming again to Ku Mark Surk, he had met there a small reindeer caravan of Yakuts bound south for Bulun and with that caravan, clad only in tattered underwear and sick almost to death, he had come across two strangers feebly expostulating with the natives against going south and almost hysterical at their inability to make themselves understood.

He spoke to them in Russian, with no better luck at communication than the Yakut reindeer drivers had had, but suddenly recalling what we had told him of our two lost boats, he enquired of them:

"*Jeannette? Americanski?*" and immediately the men had understood, nodding vigorously in assent; and writing this note, had placed it in his hands, begging him piteously:

"*Commandant! Bulun! Bulun!*"

That he understood also, but as he was bound for Jamaveloch and knew that I would be most interested in the matter, he had forthwith resumed his journey, and now, two days later, there was the message in my hands, while Nindemann and Noros no doubt were by this time in Bulun itself.

I retrieved the note from Bartlett and read it again carefully. De Long had landed, but simply "on Siberia." Where was he now? The note was blank on that. I could not tell. But evidently he was in a bad way, for Nindemann and Noros, somehow separated from their shipmates, were from Kusmah's account obviously far gone, and as for the others, that closing scrawl:

"*Reply in haste; want food and clothing*" had an ominous ring. And then my eyes fell again on "the CAPTAIN and DOCTOR and nine (9) other men."

Nine? Hastily I counted up. The captain, the doctor, Ninde,-mann, Noros and nine others—that made only thirteen! But De Long had had fourteen all told in his boat! Was nine an error? No; as if to emphasise it, the *nine* was repeated as a figure in parentheses. So already one of De Long's party had died. Sadly

I wondered who. Collins, perhaps? No, I decided; Collins had done no work on the ice to wear him down. Lee, my machinist, was most likely, I concluded. His injured hips would have made it most difficult for him to keep up and he might have had to be left behind.

But this was no time for wondering. Only Nindemann and Noros could tell me where the captain was and how to get there. And if those two men were as badly off as Kusmah said, they might both soon die, taking their secret with them. The Lena Delta was large, over 5,000 square miles in area, and from bitter experience I knew now how difficult it was to find one's way amidst its myriad islands, swamps, and freezing streams. And as for charts, there were none worthy of the name—Petermann's, which had nearly led my party to starvation, was worse than useless. I shivered as I thought of that. De Long, relying on that same Petermann chart, had intended to land near Barkin! Barkin and the north coast of the delta thereabouts, were not only uninhabited and a hopeless stretch of barren tundra, but a hundred miles farther north and by so much farther removed even from such slight shelter as we had providentially encountered at Jamaveloch! De Long and his party must be in fearful straits!

"Kusmah!" I said sharply. "Return with me to Bulun at once! Get your dogs! We start right away!"

But Kusmah demurred, objecting that his dog team was completely worn out and could not travel the ice again without several days' rest. On investigating his dogs, this proved to be true, so getting hold promptly of Nicolai Chagra, I insisted vigorously that he provide immediately from somewhere another team, if it stripped the village of its last dog.

Chagra was willing enough, but it took him all night to scrape up the necessary dogs, and not till next morning, October 30 (a day which later became indelibly burned into my memory), behind a team of eleven dogs driven by my original Yakut pilot, did I set off in a temperature twenty degrees below zero for Bulun, where from Nindemann and Noros I hoped to learn of De Long's whereabouts. Two days later, after hard labour by the dogs through deep snow

w

and over broken ice, I was at Ku Mark Surk, where I changed my worn-out dog team for a reindeer sledge, and with that made the last sixty miles southward up the frozen Lena to Bulun, arriving on the evening of November 2.

I promptly enquired my way to the hut where were lodged Ninde- mann and Noros, and in mingled fear and hope hurried there. What was I going to hear of my captain, of Dr. Ambler, about my other shipmates? Which one of them was already dead, what chance had I of rescuing the survivors? With my heart pounding violently, I pushed open the door of the hut.

CHAPTER XXXIII

I N the smoky light of the rude interior of that Yakut hut, I saw at first only Louis Noros, clothed in ragged woollen underwear, bending over a rough table, sawing away with his sheath-knife on a loaf of hard black bread, while in a corner by themselves a number of Yakuts were busy over the fire on their own supper. Noros glanced up on my entrance, looked at me vacantly, and then resumed his hacking at the hard bread. I waited a moment to see if he might recognise me, but as he did not, I advanced, stretched out my hand and said:

"Hello, Noros! Don't you know me?"

Startled at being addressed in English Noros dropped his knife, peered intently in my face, and then fell on my neck sobbing:

"My God, Mr. Melville, are *you* alive?"

At this outburst, through the smoky room I saw Nindemann suddenly lift himself on one elbow from a rough couch at the side and cry out brokenly:

"Mr. Melville! We thought you were dead! That all hands on the *Jeannette* were dead except me and Noros! Louis and me thought we were the only survivors—we were sure the whale-boat's were all drowned as well as the second cutter's!"

Bending over Nindemann, too far gone to lift himself, while Noros clung round my shoulders, I wept with them.

"No, boys," I said gently, "the whole whale-boat's crew is safe. And they're all overjoyed to know that you are too. But who died in your boat, and where, for God's sake, are the skipper and the rest of your boat's crew? I'll go for them right away."

"No use! They must be all gone by now!" sobbed out Nindemann feebly. "Over three weeks ago, October 9, the captain sent me and Louis south to look for help, and they were nearly dead then; no food for seven days and everybody frozen bad. We struggled to the south along the river and were no more able

311

even to crawl and nearly dead ourselves when the natives found us twelve days after and carried us here." Nindemann's choking voice broke hysterically. "Mr. Melville, we didn't want to come here, we wanted them to take us back! But we couldn't make anybody understand about the captain. And he was dying then. Now it's too late!" and falling back on his wooden couch, Nindemann wept like a baby in my lap.

"Where are they now?" I asked sadly. "I'll find them! Tell me; what happened, boys?" and as I listened, the tears streamed down my roughened cheeks as between them, Noros and Nindemann poured out the story of the first cutter and its crew.

CHAPTER XXXIV

BEFORE the steadily-rising Arctic gale, the *Jeannette's* three boats in broken formation were scattering in the storm. Dismayed at this sight, De Long, who had the only navigating outfit in the flotilla and in addition was carrying all Chipp's meagre food supply, rose in the stern-sheets of his cutter and waved vigorously to the other boats to get back in position astern of him. But seeing the whale-boat nearly swamp attempting to drop back, he signalled her on.

Taking a second reef in his own sail to deaden still further his speed, De Long continued waving to Chipp, hopeful at least of getting him close enough aboard to toss over his can of pemmican before in the storm and the night, he lost him to view. Badly flooded himself by oncoming seas, he nevertheless held back, till Chipp and his boat, suddenly engulfed in the waves, disappeared for ever from sight.

Sadly then, De Long shook out one reef and picking up headway, stood away dead before the wind, heading south-west for Barkin. Blond and bearded Erichsen, tall and brawny, a sailor from his childhood in far-off Denmark and in stature a royal Dane indeed, the best seaman in the boat, steered. Crowded into the cockpit before him were De Long, Ambler, and Collins, while forward of them on each side of the boat, backs to the weather-cloths holding them up against the sea, were the rest of the crew—Nindemann, the quartermaster, tending the sheet, Lee, Kaack, Noros, Görtz, Dressler, Iversen, Alexey, Ah Sam, and Boyd. Jammed under the thwarts, practically filling all the spaces there were the sledge, the tin cases containing the *Jeannette's* records, the navigating gear, the silken ensign in its oilskin case, the rifles, tents, sleeping-bags, cooking-pots, and a few cans of pemmican, with Snoozer, the last Eskimo dog, crouching on the sleeping-bags and whimpering piteously as the spray soaked him.

The heavily-laden first cutter, only twenty feet long but wider of beam than any other of the *Jeannette's* boats and with all that ballast on her bottom, therefore more stable and more resistant to capsizing than either of the other two, lumbered on before the wind, pitching heavily as the curling seas swept up under her square stern, and yawing badly in spite of all that Erichsen at the tiller could do to hold her on her course. Darkness fell, the seas grew worse, the crew bailed steadily.

Twice the boat yawed suddenly and the sail jibbed violently, straining the mast, but each time Erichsen managed to catch her and the yard and sail were again squared and the boat stood on with the wind screaming by and the merciless seas breaking heavily over the stern, soaking Erichsen continuously and spraying everyone else with freezing water.

For an hour the boat stood on before the storm with the water coming in over both sides and the stern, while her crew bailed vigorously to keep up with it. And then, riding on the crest of a tremendous wave roaring up astern, came disaster. The boat took a bad yaw as the sea struck, the stern swung off to port with the crest. Immediately the sail, caught flat aback by the wind, jibbed over and the yard banged viciously round to leeward, heeling the boat sharply down on her port side and riding the lee gunwale completely under! Instantly solid water came pouring in over the submerged rail. In another split second, the half-capsized cutter would have been bottom up with her crew spilled into the raging seas, had not at that instant the mast, already weakened by the previous jibs, broken clean off, and with the flapping sail shot overboard, momentarily relieving the fatal strain!

For one horrible second, the listing boat hung with her gunwale under, poised uncertainly between going completely over and rolling back, while her agonised crew, clinging desperately to the thwarts to avoid being tossed out, felt cold death in the form of the inrushing water lapping round their bodies! Then slowly, very slowly, under the influence of the heavy ballast jammed along the bottom boards, the dismasted cutter rolled back on an even keel, awash to the thwarts and so deep in the water that her gunwales barely showed above the foaming surface!

"Bail!" roared De Long. "All hands! Bail!"

A waterlogged wreck, the first cutter lay broadside in the trough of the sea, with every man in her buried in salt water to his waist, frenziedly bailing to regain a semblance of buoyancy before the next wave swept over her side and finished her. Fortunately, at that instant, the broken mast and the ballooning sail, dragging alongside by the halliards streaming over the bow, caught the water, began acting as a sea-anchor, and the startled men in the boat, too busy bailing to lift a hand for any other purpose, saw in amazement their submerged cutter swing slowly round in the trough into the wind and sluggishly rise head on to the next crest, heaving herself to!

Had even another moderate sea swept up at this moment, the boat would unquestionably have finished filling and foundered, but by some freak of the storm, only a succession of lazy billows came rolling by until with the boat half-emptied and higher in the water, De Long could get out some oars to hold her steady the while he sent Görtz and Kaack racing forward to clear away the wreckage.

Holding his cutter head on with oars and rudder while he finished bailing and dragged in the impromptu sea-anchor by the halliards, the captain hastily made a drag of his sail only and an empty water breaker to hold it up, and then rode the gale to that, taking in the oars and raising the weather-cloths again, while Erichsen, still clinging to the tiller, steered into the wind and the rest of the crew bailed to keep afloat.

For the men crowded in the boat, it was a night of utter misery and terror, wet through, freezing in the gale, tossing madly in the cutter, and with Collins slumped in the cockpit too weak or too heedless to reach the rail, violently and continuously seasick to add the final touch.

At midnight their sea-anchor carried suddenly away and with it went the sail. Instantly, out went a pair of oars to hold her up, while another drag, made of the broken mast and the rest of the oars, with the expedition's solitary pickaxe hung to it to hold it down, was sent out over the bow. This proved a poor substitute for the sail, for having insufficient surface, it failed to catch the water properly and rode off the cutter's beam, instead of ahead, with the result that the boat, no longer bows-on to the waves, wallowed in the troughs and rolled horribly, making water worse than ever.

After thirty-six hours of this torture, the gale finally abated, and with only a fresh breeze and a heavy sea still running, De Long prepared again to get under way, but he had no sail. Nindemann searched the boat for substitutes; out of a hammock and the sledge cover, sewed together by Görtz and Kaack, Nindemann provided a jury sail. The drag was hauled in to recover the mast; with a chisel, Nindemann refitted the broken end of the mast to its step, re-rigged it, and soon with the two insignificant bits of canvas spread at the yard, the first cutter resumed her journey for the Lena Delta, making hardly one knot through the water, and because the breeze had now hauled to the south, unable to sail closer to the wind than a course due west instead of the desired south-west.

But what the course should be to make Cape Barkin and where the boat was, God alone knew. It was impossible to get a sight, and even had it been, De Long's hands were so badly frozen he could not work his sextant. So willy-nilly, the boat went west for two days with De Long and his frozen crew, barely crawling along under the tiny jury-rig, till early on the morning of September 16, having been four days at sea since leaving Semenovski Island, the wind failed altogether, and in a dead calm sea, De Long ordered out the oars and headed due south, feeling that he had made more than enough westing.

By this time, from long-continued watchfulness and exposure, both De Long's hands and feet were so badly frozen and had swollen to such size that he was wholly unable to move himself. Tenderly, Dr. Ambler got out his soaked sleeping-bag, and helped by Görtz, slipped the captain into it and then propped him up in the stern-sheets so he could see to manœuvre the boat.

After a few hours of rowing south the water began to shoal rapidly and the cutter ran into a skim of young ice, through which it broke its way. Soon low-lying land was sighted to the south, undoubtedly some part of the northern side of the Lena Delta. With redoubled energy the men heaved with their cracked and bleeding hands at the oars, driving through thickening ice toward the coast. A little to starboard an open lead in the young ice was sighted, seemingly running inshore toward a river mouth, and into this lead the boat was rammed through the intervening ice, keeping on in this open water till at about nine a.m., still more than a mile offshore, the boat grounded solidly in less than two feet of water, with new ice freezing constantly all about her in the bitter cold.

After a fruitless effort to get inshore through the invisible shoals, De Long tried to work out again to the northward, hoping then to go farther west and perhaps find a better channel into the river, which so far as could be judged from the width between the headlands, seemed to be one of the main northern mouths of the Lena. But the thickening ice had closed in behind, and stuck fast in the hidden shoal, the boat could be moved neither ahead nor astern with the oars.

De Long, after a futile effort to push out, using the oars as poles, became desperate.

"All hands over the side to lighten the boat!" he ordered. "We'll push her off!"

Silently, all except the helmsman, the men started to obey, but first began to remove their wet boots, not wishing to fill them with mud. Off came the worn and leaking footgear, exposing to view badly-swollen feet, many already black with frost-bite and with blisters breaking as the skin, stuck to the boots, tore away from the frozen flesh. Dr. Ambler took one swift glance at them, then leaning over the helpless captain, whispered in his ear. De Long bent forward, looked himself, then said:

"Belay going over the side, men. Put on your boots. We'll try shoving her off again with the oars instead."

But the enfeebled seamen had little luck. An all-day struggle with the shallow water moved the boat hardly a hundred yards, and night fell on an exhausted boat crew, caught amidst ice and shoals, unable either to get the cutter ashore or get it to sea.

Once more they spent a cheerless night in the cramped boat,

tantalised by that unapproachable shore a mile and a half away, unable to sleep, wet, freezing, and thirsty on the crowded thwarts.

At daybreak, they tried again. Managing to get free of the ice and the mud, they made a few yards, only to ground on another shoal. Getting clear of that, the ice soon blocked them. It made little difference which way they headed, north or south, east or west, shoals and young ice were everywhere. Bitterly De Long looked from his heavy cutter and his fast-fading men across a mile and a half of thin ice, strong enough to block the boat, too weak to sustain a man, toward the low coast of Siberia. It was three o'clock in the afternoon. He would never get the boat free—eight hours of labour to-day on top of all of yesterday and no progress made either toward shore or toward sea and nothing to look forward to now except another terrible night in the boat in the fierce cold.

De Long made up his mind. Regardless of their condition, they must abandon the cutter, wade ashore. He still had two hours of daylight in which to work, and despite frost-bitten feet, there was no alternative; into that icy water they must plunge. But three of the men, Boyd, Erichsen, and himself, hardly able to stand without toppling headlong, could never make that mile and a half wading through ice and shoals to the land. They would have to get the boat closer first.

"Except the sick and the doctor, all hands over the side! We're going to abandon the boat and wade ashore! Keep your boots on this time, men!"

Slowly the rest of the crew crawled over the side into the water, finding it knee-deep. Leaving in the boat only the four men and Snoozer, and taking as heavy a load on his back as each could carry, the crew set out for shore, Nindemann first to break a path through the half-inch ice, then in succession Kaack, Görtz, Iversen, Lee, Dressler, Collins, Alexey, Noros, and finally Ah Sam, whose feet were in such bad shape that not to impede the others he was ordered to go last. It was hard work, especially for Nindemann, smashing ahead through the ice, with the chilly water changing irregularly in depth from knee-deep to over his waist, sinking unexpectedly into mud-holes from which he could hardly drag his feet, and all the while pounding away at the sheet ice with hips and thighs, unable to use his arms because of the load on his shoulders.

Finally, the panting quartermaster reached the shore, a low and swampy slope. Behind him trudged the others, and thankfully coming up out of the sea, squeezing mud and slush from their boots at every step, they dumped their loads on the beach. Siberia at last! A feeble cheer burst from husky throats and cracking lips.

But looking round at that dismal shore, covered with snow, bare of all vegetation, utterly desolate and devoid of any trace of human habitation then or ever, it is doubtful that there could have been found on earth any group of human beings save only these few who had gone through hell on ice to reach that shore, who would not have cursed instead of cheered at setting foot on that bleak tundra.

CHAPTER XXXV

"COME on, boys; we go back now for the rest of the load and the captain," ordered Nindemann, who with his rating of quartermaster was senior in the group ashore. "Shake a leg; we got lots to do before dark yet."

"Yah," said Iversen, plunging back in the sea, "frozen feet ban yust too bad for any man. Ay tank ve better get it done qvick before yet it gets colder!"

One by one, the men slipped back into the narrow lane broken through the ice after Iversen and stolidly plodded off in the water toward the distant boat, till only Ah Sam and Collins were left.

"Shake it up there, you fellers; we ain't got much time," growled Nindemann.

"I'm ashore now and I'm going to stay ashore!" snarled Collins. "Do you think you're going to get me a mile out in the ocean again wading through that mud and ice to drag in the captain and the dog? Well, you're not! I might for sick men, but not for them!"

"But the captain is sick! He can't walk!" protested Nindemann. "And besides, there's all our food and the records to carry in yet!"

"Well, he can swim then for all I care!" replied Collins defiantly. "And as for those records, carry 'em ashore yourself. I won't; I didn't ship to be treated like a common sailor, and you can't make me!"

"Suit yourself," mumbled Nindemann uncertainly, for Collins was after all an officer. He turned to the Chinese cook. "Get under way there, Ah Sam."

Poor Ah Sam, with his feet benumbed from constant immersion while bailing, staggered toward the water, then collapsed in the mud, unable to rise. The quartermaster dragged the inert China-man back on the beach and deposited him at Collins's feet.

"Get me a fire started here then, Collins, and see maybe if you can thaw him out before I get back," ordered Nindemann. "I'm going for the captain," and he plunged into the icy seas.

"Where's Ah Sam and Mr. Collins?" asked De Long anxiously when Nindemann, much behind the others, returned to the boat. "Anything wrong?"

"They're all played out," lied Nindemann glibly. "So I left 'em to make a fire for us when we got back ashore."

"Poor devils!" muttered the captain sympathetically. "You should have left somebody with 'em, Nindemann."

"Oh, they'll be all right soon," Nindemann assured him. "Besides, I needed here everybody," and in that he was right enough, for it took three trips with the seamen slithering through mud and water to get all the baggage ashore through that mile and a half of broken ice, and it was completely dark when Nindemann at last gathered what crew he had left round the lightened boat and attempted to work it ashore. But even lightened to the utmost, with nothing but the three incapacitated men and the doctor left in it, half a mile from the beach it stuck finally in the mud and they could get it no farther inshore. The wind freshened, bringing a blinding snow-storm, blotting out everything. How to get the invalids ashore was now a problem; in the slimy and uneven footing through the shoal water they couldn't safely be carried. There being no other way, one after the other, Boyd, Erichsen, and De Long were lifted over the side of the cutter by Dr. Ambler, and stood up in the knee-deep water on their frozen legs. Then, each held from falling by a seaman alongside, the three sufferers partly stumbled, were partly dragged in the falling snow across that last half-mile through the broken lane of ice to the shore, while following them, Alexey, the Indian hunter, with Snoozer over his shoulders, brought up the procession, finally emptying the first cutter of its passengers.

It was eight at night and bitterly cold when De Long and his companions, ashore at last on the desolate beach, joined his forlorn seamen crowding round the fire which Collins had started and which Noros and Görtz soon built up with driftwood into huge proportions—the first bit of warmth the water-soaked men had felt in five days of frigid Arctic weather. But it was of little comfort; beneath the snow the ground was wet, and as the fire blazed up, it further softened the beach round about it, so the men trying to dry themselves before the fire soon found instead that they were sinking into the mushy tundra to their knees.

"It's no use, men. We might as well turn in. Pitch the tents," ordered De Long wearily, and soon the two tents were erected, a little shelter at least from the cutting wind. On the soft and snow-covered ground inside them the wretched mariners stretched themselves out full length, for the first time since leaving Semenovski Island, able at least to turn in lying down.

More like stiffening corpses than sleepers, the exhausted men sprawled out in the snow and soon as the driftwood fire died away, darkness and falling snow enveloped the silent tents, while only the whistling of the chilling wind kept watch over De Long and his thirteen worn companions, stretched out at last on Siberian soil, victors in a heroic retreat over ice and ocean to which the long annals of the sea, whether in the tropics or round about the poles, offers no parallel.

Morning dawned; it snowed intermittently. Crawling from his tent, De Long looked about. Near-by to the westward, flowing north to discharge into the sea, was a wide river. From the chart, this was evidently the River Osoktah, the main northern mouth of the Lena, and close at hand should be Sagastyr, with its signal tower and a busy trading village. But, with a sinking heart, De Long, looking over the snow-covered tundra, saw that every evidence of civilisation shown on his chart was completely missing—no signal tower, no village, no signs of river traffic on the Lena, not even the slightest sign of roving hunters! Petermann's vivid description of traffic and of settlements at the Lena mouth were only the idle dreams of an unreliable geographer, as unreal as the Grecian myths of marvellous Atlantis to be found just beyond the Pillars of Hercules!

On rescue at this point De Long had based all his plans, figured his food supply, and savagely driven himself and his men far beyond human endurance to get here. And now at this long-sought goal, plainly evident to all hands, was nothing but disillusion and despair!

Hobbling about him, trying to dry themselves before a new fire, were his worn and crippled companions, all hope gone from their haggard faces, all strength gone from their frozen bodies, through bleared and sunken eyes, watching him apathetically. De Long beckoned to Ambler.

"Do what you can for the men's feet to-day, doctor, while I sort over our stores. There's no hope of assistance on the coast. We may as well look this situation in the face, and prepare ourselves to walk inland to the nearest settlement."

"And where will that be?" asked the surgeon anxiously.

"At Ku Mark Surk, ninety-five miles to the southward," replied De Long.

"Ninety-five miles!" repeated Ambler in dismay. "Why, some of these men can't walk even a mile!"

"They've got to now," answered the captain grimly. "Get to work on our feet, doctor. Our lives depend on them now. To-morrow they've *got* to carry us along!"

"Aye, aye, sir. But ninety-five miles over this tundra! In our state now, it's worse than that drag over the pack. We'll never get there!"

"Some of us may, and we'll all try. It's our last chance. And it's up to you, doctor. See what you can do to save our feet!"

All day on one man after another, Surgeon Ambler worked with lint, with vaseline, and with his scalpel, opening blisters, cutting away dead skin and flesh, gently massaging frozen feet and legs to restore circulation, and finally bandaging up. When evening fell, De Long, Boyd, and Ah Sam could hobble again. Even Erichsen, whom the long motionless hours at the tiller during the storm at sea had left with a far worse frost-bite than anyone, whose two feet, stinking with festering sores nauseated even the doctor as he worked on those horribly swollen and blistered lumps from which protruded black and feelingless toes, claimed to be improved and able to walk a little.

While this (during a storm of snow, hail and sleet) was going on, De Long ordered a cache made on the beach of the navigating gear, most of the cooking utensils, the sleeping-bags, and other miscellaneous articles, so that the baggage to be carried was reduced to the clothes the men wore, the ship's records, four rifles and ammunition, medicine and surgical tools, blankets, tents, and their four days' food supply, consisting only of some tea and the unopened can of pemmican which should have gone to Chipp.

Leaving a written record in the cache to direct anyone who might ever come after, searching for them, on the early afternoon of September 19, the ragged seamen shouldered their burdens and dragging the expedition's records on their little sledge, set out under a bright sun over the snow-covered tundra for Ku Mark Surk, ninety-five long miles to the south over the trackless delta.

It was a forlorn scene as De Long and his men took leave of the Polar Sea which for two years had held them prisoner—to the west flowed the Lena, a broad swift stream tumbling on its swirling bosom broken floes from farther up the frozen river; to the north spread the Arctic Ocean, covered as far as eye could reach with young ice, through which, sticking up gaunt and bare, the only objects visible on its desolate surface, were the mast and the low gunwales of the abandoned cutter. To east and south lay the flat snow-covered tundra, and over this straggled the dismal caravan of the first cutter's crew—Iversen and Dressler dragging the sledge, Alexey out ahead to break a path, De Long following him with the *Jeannette's* ensign in its oilskin case slung across his back, and behind him the rest of the seamen staggering under their loads, with Lee, whose weakened hips frequently gave way under him, constantly falling in the snow, and Erichsen, Boyd, and Ah Sam hobbling painfully along at the rear.

It was terrible going, not helped much by a fifteen-minute pause every hour for rest. The snow-covered ground was swampy, with many ponds covered with thin ice and hidden under the snow, and into these pitfalls the men stumbled frequently, burying themselves to their knees in the mossy tundra beneath, and coming up with their leaking boots or moccasins filled, to plunge along again through the snow and the freezing wind, oozing a slimy mixture of mud and water from between their toes at every step.

Big Erichsen could barely even hobble, hardly able to lift one numbed foot after another. At the second step for rest, Ambler drew Nindemann aside:

"Quartermaster, can't you make a pair of crutches for Erichsen? His arms are still strong; with crutches, he'll make out better."

"Yah, doc, but with what should I make 'em?" asked Nindemann. "I ain't got tools no more."

"Don't worry over that, Nindemann," replied the doctor. "You've got a knife." He opened his medicine chest on the sledge. "Here, take my surgical saw; I guess if it'll saw bones, it'll saw wood all right," he finished grimly.

Nindemann got to work on some driftwood branches, and soon between sheath-knife and bone-saw, he had fashioned a fair enough pair of crutches, on which when the party resumed its journey, Erichsen swung along haltingly behind the crippled Ah Sam.

But for the worn and burdened seamen, progress was still snail-like. After another faltering advance, De Long halted the party and deciding to lighten up still farther, sent back Nindemann and two other seamen with one tent, all the log-books, the spy-glass, and two tins of alcohol to stow them with the abandoned gear in the cache at the beach. This left to be carried or dragged by the men only De Long's private journals as a record of the expedition, one tent, some alcohol and medicines, the rifles, a cooking-pot, and what little food they still had, together with the silk flag which De Long himself bore along.

The second day thus, the party staggered on four miles more to the south. The going got worse, the straggling procession length-ened out in the snow. A brief pause to rest, and all hands once more got under way except Nindemann, whose load, chafing his shoulders, stayed behind to readjust it while the others started off through the snow. Having eased the fastenings of his pack as well as possible, the wearied quartermaster struggled to his feet and was hurrying forward to catch up with his mates when unexpectedly he stumbled over what as he fell he thought at first was a log half-hidden in the snowy path, but which he quickly saw to be Erichsen,

x

prone on his face, while near-by, tossed into a drift, were his crude crutches!

With a thumping heart, Nindemann feverishly rolled his shipmate over on his back expecting to have to revive him, only to find instead Erichsen's snow-flecked blue eyes staring bitterly at him, and Erichsen's broken voice rising in a curse:

"Go avay, damn you! Ay vant yust to die here in peace!"

"Get up, Hans!" pleaded Nindemann. "You're not going to die; nobody is. Here's your crutches. Come along! I'll help you!"

Erichsen only shook his head, his eyes rolling in anguish.

"No use, Nindemann, my feet ban all gone! Even if you can go so far as Moscow, Ay tal you, Ay cannot go one step more! Go on! Let me die!" and with a convulsive effort of his huge body, he twisted himself face down again and clawing feebly with his fingers, tried to bury himself completely in the snow.

Frightened, Nindemann jerked erect and shouted down the trail:

"Captain! Hey, captain! Come back!" but so far off were all hands now that no one turned. Leaving his silent shipmate in the drift, the quartermaster, going as fast as the broken path allowed, hurried after them, shouting occasionally, till half a mile along he finally attracted De Long's attention and stopped him till he could catch up, when he told the captain of Erichsen's plight.

De Long gritted his teeth.

"Keep ahead, Nindemann, till you come to driftwood, then build a fire quick and camp," ordered De Long briefly. "Come on, doctor; we'll go back for Erichsen!"

Back rushed De Long and Ambler; still buried in the snow, as Nindemann had left him, they found the prostrate Erichsen. With some difficulty, Ambler turned him over, while De Long pulled his crutches out from the deep snow alongside. The doctor took the broken seaman by both shoulders and started to lift him.

"Let go me, doc," begged Erichsen, "it ban no use any more to help. My legs ban killing me. Ay vant now only to die qvick! Go avay!"

"*Get up, Erichsen!*" ordered De Long in a voice cold as steel. "Here's your crutches; take 'em and get going down that road!

Do you think I'm going to leave you now? Get under way! And when you can't hobble, I'll drag you! Up now, before I jerk you up!"

For a moment, Erichsen, lying in the snow, stared dumbly into the captain's inflexible eyes, then his habit of obedience conquered his suffering. Slowly he pushed himself into a sitting position and without another word reached for the crutches. With Ambler's assistance, he rose to his feet and then with both De Long and the doctor behind him to see that he did not again lie down, he hobbled off down the path, each step undoubtedly an agony to him as his bleeding and tortured feet came down in the snow. And so, slowly and painfully, they covered the last mile into the camp, where a roaring driftwood fire and a scanty supper of cold pemmican and tea awaited them.

Before the fire, all hands steamed in front while they froze behind, and then stretched out on driftwood logs for a bed, hauled their sole remaining tent flat over the fourteen of them and turned in. But between sharp winds, bitter cold, and falling snow, it was a fearful night for the fourteen sufferers, shaking and shivering beneath the thin canvas, and no one slept.

Through snow and fog again the party struggled southward along the river bank next day, with Boyd and Ah Sam both improved, and even Erichsen, the captain's stern voice still ringing in his ears, doing a little better on his crutches. But with only two days' slim rations of pemmican left, and with each day's progress hardly a scant five miles over the snowy tundra, the chances of making the remaining eighty miles to Ku Mark Surk began to fade.

In the middle of the third afternoon, the party came to two abandoned wood huts by the river-side, the first evidence of habitation they had met in the Lena Delta, and gladly all hands entered. Inside the huts, reasonably sheltered for the first time in weeks from cold, from wind, and from snow, and with plenty of driftwood about so they could warm themselves at last, the men stripped off their soaked and ragged furs and stood about naked while their clothes dried before the hurriedly-built fires.

Dressed again, and with a tiny portion of pemmican and some

hot tea for supper, the exhausted travellers threw themselves on the dirt floor, at last to catch some sleep inside a human habitation, primitive even though it was. No one any longer had a sleeping-bag; only the patched and ragged remnants of the fur and cloth garments and the long since worn-out boots in which three months before they had started the terrible journey over the ice from the sunken *Jeannette* remained to them. But at least there was a tight roof and solid walls about them and it was enough. In a few minutes, at four o'clock in the afternoon, thankful beyond description for so much shelter, all hands were sound asleep.

But there was one exception. Shelter or no shelter, Erichsen, suffering the agonies of the damned from his mortifying feet, only tossed and moaned, waking the doctor. Rousing Nindemann to help him, the surgeon seated the suffering seaman on a log before the fire, got his instruments and medicines, and then, while Nindemann held the patient erect on the log, gently proceeded to un-bandage his left foot, the worst one.

As the last turn of the bandage came off, Nindemann anxiously watching, saw to his horror, all the flesh, dead and putrid, drop away from the ball of the foot, exposing tendons and bones. Startled, he closed his eyes, repressed a groan. But Ambler said nothing; only the slight compression of his lips indicated his despair. There was nothing medical skill could do. Quietly smearing a fresh bandage with vaseline, he carefully bound up the foot again and put back Erichsen's stocking and his boot.

"All done, Erichsen," he said reassuringly; "you can turn in now," and gathering up his equipment, Ambler, his heart torn by poor Erichsen's condition, hurriedly stretched himself out in the hut as far away as he could get lest his patient should start to question him.

But Erichsen was not wholly ignorant of what had happened. Turning to Nindemann on the log beside him, he asked:

"Do you know much about frostbites?"

"Yah, Hans," replied Nindemann, "at the first coming on, the flesh turns blue and then it gets black."

The big Dane nodded, continued sadly:

"Ven doc took off the bandage, Ay saw somet'ing drop from unter my foot. You saw it too, Nindemann. Yah?"

Nindemann, with one arm about his suffering shipmate to keep him erect, looked him squarely in the eye, and putting all the conviction into his voice that he could muster, he lied heroically:

"No, Hans, there was nothing. You must be dreaming things."

"Don't try to fool me, qvartermaster; Ay tal you Ay saw it und so did you." Mournfully he gazed at his shabby boot, then sadly shook his head. "Ay hope you get home yet, Nindemann, but vit me, it ban all done. Stretch me out now; you must sleep."

But it being still early in the evening, after a brief nap, De Long sent Alexey and Nindemann out with rifles to hunt, the while the others rested and he took stock of the situation.

Long and earnestly, as the two hunters trudged outside through the snow looking for game, the captain pondered. His recent chart, based on Petermann's reports and descriptions of the villages on the delta itself, he now knew was worthless; only in the old Russian chart showing Ku Mark Surk at the head of the delta and Bulun beyond could he put any faith. But with the nearest of these over eighty miles distant, it was hopeless to expect that his crawling party, making at best five miles a day, could ever get through on the two day's pemmican still left. And without food to sustain them on the way, the outside temperature, hovering around zero, would of itself in a few more nights in the open like the preceding one, quickly make an end of them. There seemed nothing for it except to stay in the huts where at least they had shelter and warmth and stretch to the utmost their few pounds of pemmican, eked out by poor Snoozer as a last resort, the while he sent two men ahead on a forced march to Ku Mark Surk in the thin hope that he might keep his starving men alive till they returned with aid, in two weeks at the soonest if they found the travelling good, longer if they did not.

What alternatives were there? He considered them. Erichsen, Lee, Boyd, and Ah Sam were his drags on progress, especially the two former. If he left these two, the others might easily double their speed of travel and reach Ku Mark Surk and safety in possibly a

week. But it would take at least a second week to get help back
to his abandoned men. How could two helpless cripples without
food, hardly able to crawl outside to gather wood to warm themselves, stay alive for two long weeks, perhaps more? They would
soon, hopeless in the feeling that they were deserted, both lie down
and die. As it was, only his constant driving, his apparently soul-
less harshness, and the lash of his stinging commands, kept them
hobbling weakly along.

Could he abandon them? Dispassionately he tried to consider it.
On one hand, a far better chance for life to twelve men, certain
death for two. On the other hand, the strong probability that all
would perish in that hut before relief arrived. Going on, leaving
his cripples behind, looked logical. But De Long shook his head.
While he lived, he could abandon nobody to the loneliness of that
Arctic waste, least of all the heroic Erichsen, who unrelieved through
that terrible night in the boat, had clung to the tiller, safely steering
them all through the gale, and now in the agony of his decaying
feet, was uncomplainingly paying the penalty of his steadfastness.
With a sigh, the captain decided to stay on in the hut, while he sent
ahead for help. Who should go? Running over in his mind the
physical condition of his men, he decided on Surgeon Ambler and
Nindemann, the two he felt who were most likely to get through.

At six o'clock, Nindemann returned, empty-handed except for a
dead gull he had found. Eagerly, the hungry seamen, roused by
Nindemann's entrance, crowded round while Ah Sam plucked the
gull, only to discover that the carcass had long since rotted. Sadly
it was thrown away, and the disappointed sailors once more turned
in. Alexey still was missing, but no fears were felt for him, and
quickly, without exception now, the exhausted company sank into
deep slumber.

About nine o'clock came a knock on the door of the hut and
Alexey's voice rang out:

"All sleep here?"

Immediately, sleeping heads lifted here and there over the floor
as the door flew back and Alexey cried proudly:

"Captain! I shoot two reindeer!" and in staggered the snow-covered hunter bearing on his back the hindquarter of a doe.

"Well done, Alexey!" shouted the captain, leaping to his feet and kissing the startled Indian, while all about men sprang up, almost smothering the beaming Alexey in handclasps and in clumsy hugs. Immediately sleep was forgotten, the fires poked up, and that haunch of venison, cut in chunks, was roasting on a dozen sticks. Each man got a pound and a half; most of them, long before their meat was hardly more than seared before the fire, were gorging themselves on the raw flesh! With startling rapidity it disappeared, and hungrily his men looked toward the bloody remnant of that haunch, but De Long, stowing it behind him in the hut, shook his head and ended the feast, leaving the party no option but to return to sleep, while only Snoozer, still gnawing wolfishly at the shank bone, remained awake.

That changed De Long's plans. Issuing only a very scanty ration of pemmican for breakfast, he sent Alexey and six men out in the morning to get the deer, while he concluded to spend that day and the next in the hut, recuperating the sick, and then with his two days' supply of pemmican still intact and the remainder of the two does for food on the journey, push on southward with all hands.

And so they did. Warmed by soup made of the reindeer bones, fortified by deer-meat, and rested by two days' inaction in the hut, the party set out hopefully on September 24 with twenty pounds of pemmican and fifty-four pounds of venison still left for food for fourteen men and their dog, leaving a note and the captain's Winchester rifle (for which there was no longer any ammunition) as a record behind them.

They tramped along the east bank of the river for three miles, resting hourly and making poor progress. Looking hopelessly at the broad stream still flowing unfrozen past him, De Long sighed for his abandoned cutter, in which here with oars and sail, they might make fine progress even against the current. But the cutter was gone and wishes would do no good. However, they might perhaps make a raft and sail or pole that upstream, at least relieving their feet. So stopping the party, De Long turned all hands to gathering logs, out of which, using the sledge lashing for a fastening, a crude raft was finally fashioned at the cost of eight hours' strenuous toil, on which at five p.m. they attempted to embark. But the river had ebbed meanwhile, and in spite of an hour's battle, it was impossible to get the grounded side of the raft afloat. In deep disgust, amid the suppressed curses of all hands at the result (and especially of Nindemann, who had done most of the work), the raft was abandoned, the loads picked up again, and the men, doubly weary now, staggered away southward, again to camp for the night on the open tundra, freezing on a few logs spread in the snow for a bed, to rise next morning after no sleep at all, stiffer and sorer even than the night before.

The next day's travelling was difficult beyond words, over snow and thin ice, through which torn boots broke, to come up covered with a slushy mixture which immediately froze solid, soon making each man's feet as large and as heavy as sand-bags, a gruelling task to lift them, an endless labour to keep them reasonably cleared.

By some miscalculation either in issue or in original weighing, but eight pounds of deer-meat was found remaining, all of which went for dinner. An afternoon of heart-breaking travel over an ice-coated bluff from which the piercing wind had cleared all snow, leaving it slippery as glass, brought them at night to a dilapidated hut, filthy in its interior, but nevertheless the freezing seamen, taking it for a god-send, stretched themselves promptly in the dirt inside, unutterably grateful for the shelter. A scant portion of pemmican passed for supper. With only three similar rations apiece left as the total food supply, the toil-worn men turned in, grumbling audibly for the immediate issue of the remnant of the pemmican and De Long began to fear open rebellion.

Day broke inauspiciously. Before them, blocking the way, was a swift side stream, too deep to ford, with ice too thin to walk upon. De Long, after examining all possibilities of crossing, ordered Nindemann to build a raft to ferry over on, and Nindemann, tired, hungry and bitter over the fiasco attending the raft of a few days before, went grumblingly at it. While he and his shipmates struggled with the logs and the single line they had for a lashing,

De Long, silently ignoring the none too well hidden signs of growing disaffection, went back to the hut. Outside the door, Ambler
met him, pulled him aside:

"Erichsen's condition is getting desperate, captain. Both feet are
worse! another couple of days and nothing in God's world can keep
him on them."

"All right, doctor. We'll do the best we can," said De Long
resolutely. "Keep him going to the last minute, then we'll drag him.
Meanwhile, I'd better keep an eye on the work on that raft."

By ten a.m., the raft was done, a crazy affair and not very large,
due to the lack of sufficient lashings. With Collins, Alexey and Lee
as passengers, and Nindemann and Kaack as ferrymen, it started
over, amid voluble cursing promptly submerging all hands to their
knees. But nevertheless it got successfully over to shoal water on
the other side, where Nindemann started to look for a good landing
spot.

"Don't waste time!" shouted De Long. "Let those men wade
ashore and hurry back with that raft!"

After considerable growling, audible even to De Long on the
other shore, the passengers waded off, and the two ferrymen paddled
back. On his return, Nindemann promptly started grumbling again
about the raft.

"What's the matter, Nindemann?" asked De Long.

"The lashings are loose and there ain't enough logs to float it,"
said Nindemann sullenly.

"Well, you made the raft. Haul the lashing tighter then if it
doesn't suit you," suggested the captain.

"But I hauled it already as tight as I could," protested the irritated
quartermaster.

"That'll do!" Curtly De Long cut him short. "Get more logs if
you want them; tighten the lashings if you wish, but quit standing
there! I've had enough of your grumbling! Shake it up, now!
We've got to get on!"

Glowering at the reproof, Nindemann, his nerves finally at the
breaking-point, glared a moment at the skipper, then turned and
moved down the bank. A few steps off, facing the next gang of

men waiting to cross on the raft, the stocky quartermaster clenched his fists, swung them wildly in the captain's direction and shouted:

"I would sooner be along with the devil than be along with you! I wish I was in hell, or somewhere else than here, by Jesus Christ!"

Quietly, De Long looked from the little knot of men on the raft to Nindemann's circling fists, then in an icy voice, he ordered:

"Nindemann! Come back here!"

Slowly the infuriated quartermaster approached his captain, to find a pair of cold blue eyes drilling into him.

"So you'd sooner be shipmates with the devil than with me, eh? You'll find yourself in hell quick enough if you don't do what I say! What's the matter now?"

Nindemann quailed, his mutinous passion suddenly chilled before that frigid gaze.

"Nothing at all, sir," he mumbled weakly.

"Another word from you and I'll have you court-martialled!" said De Long coldly. "Now get up into that hut and consider yourself under arrest until I send for you!"

"Very well, sir," answered Nindemann, and meekly he scrambled up the bluff to the hut, while the captain looked down at the men milling round on the raft.

"Görtz! Lend Kaack a hand with those paddles! Shove off now!"

"Aye, aye, sir!" Immediately the raft started its second trip.

It was slow work. Not till three in the afternoon was the raft ready for the last load. Then sending Erichsen down first, De Long peered into the hut at Nindemann crouching before the fire.

"Pick up your traps, quartermaster, and get to work again!"

"Aye, aye, sir!" said Nindemann obediently, and hastily gathering up his load, he ran down to the raft where for the last trip he paddled over and then, dismantling the logs to recover the priceless lashings, he looked expectantly up to the captain for orders.

"Build a fire," said the skipper briefly. "We'll have dinner here and dry ourselves before moving on."

They made four miles by dark, camped in the snow, froze as usual instead of sleeping, ate a skimpy breakfast, and with but a single meal left, the party was about to break camp, when far away Nindemann spotted some reindeer approaching the river. Keeping everybody down, the captain sent Alexey and Nindemann out with rifles.

Circling three miles to get to leeward of the small herd of reindeer, the two hunters crawled cautiously along on their stomachs another quarter mile, pausing, with their very lives depending on their care, each time a deer looked in their direction, then snaking along again through the snow. At last, within a hundred yards, they stopped, picked out the two largest bucks they could see, and at a word from Alexey, fired simultaneously.

Down went the buck at which Alexey's Remington was aiming, but Nindemann's Winchester missfired and before Alexey could get in another shot, the startled herd was off. Firing, nevertheless, Alexey swung to the moving targets, but failed to hit again. Leaping up, the two men ran in to secure their prize and saw joyfully that Alexey had knocked over a fine buck, as large as both the does which he had previously shot. It took five men to drag him into camp, and there, all thought of movement suspended, the ravenous men turned to on frying deer-meat, gulping down three pounds apiece before the captain finally called a halt on eating, and ordering his crew to shoulder the remainder of the buck, provisions for three days more, they got under way again in the teeth of a driving snowstorm.

By the next afternoon, September 28, having spent the previous night again in the snow, De Long came to an empty hut on a promontory and looking off ahead, found himself trapped! On his right, running north, was the Lena; before him, running east, was another broad river branching away from it, and neither one could he ford, nor after a diligent search, find any materials about of which to make a raft. Huddled in the dirty hut, his utterly tired men sprawled out before the fire, while Alexey scouted the river to the eastward for a ford, but found none.

For three days, the ill-fated refugees were forced to remain in that hut, unable to move in any direction except back northward, while a gale outside brought heavy snow; and bitterly cursing their enforced inaction while consuming their precious provisions, they waited hopeless of movement till in the increasing cold, the river should freeze hard enough for them to cross. And meanwhile, fearing Erichsen would get lockjaw if he waited further, Dr. Ambler was forced to amputate first all his toes and then saw away a good part of the remainder of the unfortunate Dane's feet, leaving him with useless stumps on which it was hopeless to expect, even with crutches, that Hans Erichsen would ever walk again.

The captain became desperate. He cut the issue of deer-meat down to the limit, sent Alexey out in the blinding snow to hunt in one direction, Nindemann in another, and Görtz and Kaack with fish-lines to see whether the rivers which were choking off their progress, might at least yield up a few fish to eke out their provisions. But except for one gull which Alexey knocked off a pole with a rifle-ball, not a solitary bit of food did anyone get.

Meanwhile, the problem of how to move Erichsen became acute. Finding a solitary driftwood plank, six inches thick and about four feet long, Nindemann was turned to with a hatchet and the doctor's saw (which but a few hours before had been used on Erichsen's feet), to make a sledge on which to haul him, and by the night of September 30, it was done.

October 1 came and the Arctic winter descended on them in earnest. After a bitterly cold night, they issued from the hut to find the Lena apparently frozen from bank to bank. Cautiously, with the thin ice cracking ominously beneath them at each step, Alexey and Nindemann scouted a path across, then one by one, with the men widely separated, to distribute the weight, the others crawled over, last of all Erichsen on his sledge drawn by two men some distance apart hauling on a long line.

With all hands finally on the west bank without mishap, the party turned south and for three days struggled on through increasingly bitter cold, never finding any shelter, sometimes travelling on through the night, because that was less of a torture than freezing while stretched out in the snowdrifts. The delta became a maze of intersecting streams among which De Long was wholly unable to locate his position on his useless charts. And a new horror was added to their others—Erichsen became delirious and each time the shivering men halted, he raved incessantly in Danish and English, making sleep impossible even had the frigid nights otherwise permitted it. And then the food (except for tea) gave out completely, first the remaining scraps of reindeer going; finally the last hoarded bits of the pemmican (which for nearly four months they had dragged with them from the *Jeannette*) went for dinner on October 3.

Without food the party staggered on from their dinner camp, De Long praying earnestly that some game might by a miracle again cross their path. But they saw none, and weak with hunger dragged their ice-clogged feet along, skirting the thin ice on the river edge where the going was easier than on the mossy snow-covered tundra. Suddenly, De Long broke through and went into the river up to his shoulders; while he was being hauled out, Görtz plunged through to his neck and Collins was soused to his waist. A moment after they had been dragged back to the surface soaked to their skins, each was a glistening sheet of ice, with no help for it but to keep hobbling onward till evening, when still in the open, they camped on the river bank and, in the midst of a whistling gale of wind and snow they huddled round a driftwood fire where the ice-coated sufferers endeavoured vainly to thaw themselves out.

There was nothing left for food for the wan and hungry crew—except Snoozer. De Long, hoping to take at least this favourite dog back home with him, had clung tenaciously to Snoozer through thick and thin, kept him in the boat when the other dogs ran off at Bennett Island, saved him when the other seamen would have left him to starve or drown in the abandoned cutter off the Siberian coast, fed him from his vanishing store of pemmican when he had little enough to eat himself. But now, with his men starving about him desperately needing food if they were to hold a little life in their chilling bodies, sentiment and affection had to give way. Sadly he called over Boyd and Iversen, told them to take Snoozer where no one could see them, kill him, and dress the carcass.

So for supper each had a little dog-meat eaten with revulsion by everyone, but eaten. And then followed a night horrible beyond description. Erichsen's ravings mingled with the whistling of the wind; in the sub-zero blackness, the stupefied men, unsheltered from the driving snow, crouched about a fire from which they could get no warmth; in his wet and freezing garments, De Long huddled alongside Alexey to keep from freezing to death; while all about, shivering limbs and chattering teeth beat a gruesome accompaniment to Erichsen's groans as lashed to his sledge, as close to the fire as they dared put him, he alternately shrieked and moaned in delirium till finally he lapsed into a coma.

Morning came at last, to bring the unpleasant discovery that Erichsen had somehow worked off his mittens during the night and both his hands were completely frozen, through and through. The doctor set Boyd and Iversen to work chafing his fingers and palms, endeavouring to restore the circulation, but it proved hopeless. Erichsen was now totally unconscious.

Meanwhile, Alexey had spied a hut a few miles off, and after a hastily-swallowed cup of tea which constituted breakfast, the men hurriedly shouldered their burdens and dragging their unfortunate shipmate, moved off toward it, fumbling along through the driving snow and the intense cold for two hours, when, fervently thanking God for the shelter, they reached the hut and building a fire inside, proceeded to get warm for the first time in four days.

Here, after a brief prayer for the unconscious Dane, read in a broken voice by the captain, the entire party (except Alexey) sank to the floor to rest at last. Alexey refused to rest. He had shipped for the cruise, not as a seaman but as a hunter, and now with his captain and his mates urgently needing food, regardless of himself, he went out to seek it. But there was not the slightest sign of game about, and frozen worse than ever from having broken through the river ice on his hunt, the faithful Indian was at last compelled to return empty-handed.

Supper, half a pound of dog-meat apiece, and the last of the tea, was the only meal for the day, but grateful to be out of the blizzard raging round about, no one complained.

October 5 came and went, commencing in a breakfast consisting only of hot water coloured by re-used tea-leaves and ending with a supper composed of the last of the dog-meat and more hot water barely tinted with third-time-used tea-leaves. Hour by hour the men sat, crowded in the little hut gazing at Erichsen, occasionally conscious now, while his strength slowly ebbed away and his tongue babbled feebly about his far-off Denmark. Night fell, the storm howled on, the dying seaman relapsed again into a coma, and his overwrought shipmates sagged down on the dirt floor to rest.

October 6 came and in the early morning light, Erichsen died. Sadly, in the driving snow, the grief-stricken sailors gathered round a hole cut through the river ice, while broken-hearted De Long sobbed out the funeral service over the body of as brave and staunch a seaman as ever sacrificed his life to save his shipmates. And there in the Arctic wastes, where he so long had suffered, with three volleys from all the rifles in the party ringing out over the ice as a final salute, mournfully his gaunt and frozen comrades consigned Hans Erichsen, their strongest and their best man, to the Lena's waters.

CHAPTER XXXVI

WITH some old tea-leaves and two quarts of grain alcohol as their entire food supply, the thirteen survivors gloomily resumed their southward trek on October 7. The snow was deep and still falling; the weakened men ploughed through it to their waists. A little alcohol mixed in water constituted dinner; a little more of the same was served out for supper and night found them camping in the snow.

October 8, under way again over thin ice, De Long sought a trail over the wandering streams and through the multitude of islands where the spreading Lena flattened out over the low delta lands and its surface waters, churning in swirling eddies, were not yet completely frozen over. More and more frequently the faltering men paused to rest; De Long particularly, whose freezing immersion of a few days before had sadly damaged his feet, was in worse condition than anyone save Lee, whose weakening hips continually gave way, plunging him drunkenly into the drifts every other step. Badly strung out, the line of starving seamen staggered along with their captain in the rear, constantly refusing the offers of his men to relieve him of the load he carried and thus ease the way for him. When finally they halted for the night, shelterless on the bleak and open tundra, his hungry men had once again to be content with nothing more substantial to fill the aching voids in their stomachs than hot water and half an ounce of alcohol. De Long, watching them drop feebly in their tracks in the snow with Ku Mark Surk still (as he thought) over twelve miles away, concluded sadly that they could never all cover that .last stretch alive. Without the slightest chance now of getting food in the deserted delta, they would soon in their weakened condition use up the last dregs of their fading vitality and quickly freeze to death in their tracks. His only hope lay in sending a few stronger men ahead for help, while in some shelter, if they could find it now, the rest of them, fight off starvation, conserve their little remaining strength and await rescue. With that resolve, he beckoned Nindemann to his side in the snow.

"Nindemann," said the captain earnestly, "I'm sending you ahead to-morrow to get through to Ku Mark Surk for aid. It should be only twelve miles south now. You ought to do it in three days, maybe four at the most, and get back in four more. Meanwhile, we'll follow in your trail. I'll give you one of our two rifles, your share of the alcohol for food, and you can take any man in the party with you except Alexey to help you out. Alexey we must keep as a hunter. Who do you want?"

The quartermaster thought a moment, then answered:

"I'll take Noros, captain."

"Isn't Iversen better?" asked De Long anxiously. "I think he's stronger."

"No," replied Nindemann, "he's been complaining of his feet three days now."

"That's right, captain," broke in Dr. Ambler, who was alongside the skipper. "Noros is best."

"All right; Noros then. Be ready, both of you, in the morning." Stiffly De Long stretched himself out before the tiny camp-fire crackling feebly in the snow.

Morning found thirteen sombre seamen looking anxiously off over the frozen tangle of rivers and of islands to the south. Somewhere there beyond that terrible delta land lay Ku Mark Surk and life, but all about them was only the vast snow-crusted tundra, an Arctic waste of wintry desolation and the promise of slow death. Solemnly De Long shook Nindemann's hand.

"You'll do all a man can do to get us help, I know, Nindemann," he said. "God keep you safe and bring you soon again to us."

"I ain't got much hope of finding help, captain," responded the quartermaster gloomily. "It's farther maybe to Ku Mark Surk than you think."

"Well, do the best you can. If you find assistance, come back to us as quickly as possible. God knows we need it here! If you don't——" The captain's voice broke at that implication, he paused a moment, then concluded huskily: "Why, then you're still as well off as we; you see the condition we are in." He turned to Nindemann's companion, standing in the snow beside him:

"Noros, are you ready?"

"Yes, captain."

De Long looked them over. They carried nothing but one rifle, forty cartridges, and a small rubber bag with three ounces of alcohol, their share of the party's sole remaining substitute for food. Their clothes were ragged, their sealskin trousers bare of fur, their boots full of holes. The captain's eyes lingered on the toes protruding from the remnants of their footgear.

"Don't wade in the river, men. Keep on the banks," he finished gently.

There was a bustling in the little knot of men surrounding them, and Collins suddenly pushed through to confront De Long.

"I'm the *New York Herald* correspondent with this expedition," he said brusquely. "As James Gordon Bennett's representative, I demand the right to go with these men!"

De Long, surprised at the interruption, flushed slightly, then answered evenly:

"Mr. Collins, we'll settle that question with Mr. Bennett in New York. At present, getting you or anybody through as a newspaper correspondent interests me very little. And in any other capacity, just now you're only a hindrance to this expedition; you're much too weak to keep up with Nindemann. You wouldn't last five miles!" and turning his back on Collins, he gripped Noros's hand, shook it warmly, and repeated:

"Remember, Noros. Keep out of the water! That's all. Shove off now, men!"

Bending forward against the wind, Noros and Nindemann staggered away toward the south, the last forlorn hope of the eleven emaciated castaways standing in the frozen drifts behind them, cheering them as they vanished in the blinding snow.

CHAPTER XXXVII

"AND that was on October 9, Mr. Melville," sobbed Nindemann. "But Ku Mark Surk wasn't twelve miles away like captain thought; it was over seventy miles! His chart was bad, and besides every day before, he hadn't travelled so far as he guessed maybe. For ten whole days after that, Noros and me went south over terrible country, and we found to eat only one ptarmigan I shot with the rifle, and we ate up first our boot soles and then most of our sealskin pants and we froze and kept on going till even the sealskin pants was all gone and we had travelled over forty miles and still we had not come to Ku Mark Surk. And all the while we dragged ourselves along because we knew our shipmates could get no food in that country we had gone over and they were starving and the captain trusted Noros and me to get help for them.

"But after ten days we were freezing in only our underwear for clothes and we were so weak without food that we could not go on, and when we saw at last an empty hut, we crawled inside there to die, but we found in it a little rotten dried fish that looked like sawdust and tasted like it too, and we ate that, thinking maybe then we could keep on again, but the mouldy fish made us so sick with dysentery we could not even any more crawl, and we lay there three days expecting only to die soon, when at last some natives looked in that hut and found us! We would be dead there in that hut long ago if not for them!" Nindemann choked back a bitter sob and gripped my hand feebly. "We couldn't make them natives understand they had to go back north for the captain, and they brought us first to Ku Mark Surk and then here to Bulun. And now it is November 2, eleven more days even since they found us, and there is no hope for anybody any more! The captain and our shipmates must now all be dead in that snow!" And racked with sobs at the idea that somehow he had failed in the captain's trust, Nindemann wept hysterically.

343

"Perhaps they found shelter in a hut," I suggested, trying to calm him. "I'll start back right now to look, anyway."

"No use," repeated the quartermaster hopelessly. "For a long ways from where we left them, there ain't no huts, only a hundred rivers going every way, and for a man twice to find the same spot there is impossible. You ain't so strong no more. You'll only die yourself!"

I laid the weeping seaman back on his couch. Probably he was right. But so long as the faintest shred of hope existed for Captain De Long and his comrades, I must look for them.

I got the best directions I could from Noros and from Nindemann as to the route south they had travelled, where they had stopped each night, the rivers they had crossed. Taking either man with me as a guide was impossible; they could not travel. So leaving instructions for my whale-boat party that, except for Bartlett (who was to stay in Bulun to search for me if in a month I did not return), all the others on arrival there were to proceed under Lieutenant Danenhower's charge south to Yakutsk, I got a dog sledge and immediately started north. At Ku Mark Surk I met the Russian Commandant next day; he helped me with another dog team and a ten-day supply of fish. With that I proceeded northward along Nindemann's trail from Ku Mark Surk, having two native drivers and twenty-two dogs.

Through fierce November storms we pushed on down the delta, sometimes finding Nindemann's trail, often losing it. The going was slow, the cold was intense, we were frequently stopped by gales which completely blinded us and against which the dogs refused to travel, instead lying down in the snow and howling dolefully. The river began to divide as it spread out over the flat and treeless delta. One after another I searched along innumerable streams for Nindemann's trail, but in the deepening snows found no sign as we went north. Wrapped in thick furs, I nevertheless nearly froze to death on my sledge. It was inconceivable that De Long and his companions, long without food, clothed only in scanty rags, could live through such weather. But still I searched, hopeful now at least of recovering their bodies.

Our food gave out, the Yakut drivers wanted to return to Ku Mark Surk. I enquired if there were any village on the delta itself from which we might continue our search. They said there was one. On the far north-western corner of the delta on the Arctic shore, some thirty miles due west of where from Nindemann's account De Long had landed on the coast, was a small village called Tomat. I looked at my chart, a copy I had long ago made at Semenovski Island of De Long's. There was no village marked there on that chart, but knowing now the chart to be wholly unreliable, I accepted my drivers' statements as being true and ordered them to head for Tomat to replenish our food supply, intending then to pick up De Long's trail at the abandoned boat, and follow him southward from there till I came upon his party, whether alive or dead. But my drivers protested; we must turn about and return to Ku Mark Surk; without food, we would all perish on the desolate road to Tomat. Fiercely I turned on them in their native tongue.

"Head north!" I ordered savagely. "And when we have to, we'll eat the dogs! And when they're gone, by God, I'll eat you if necessary to get north to Tomat! Keep on north!"

Cowed by my threats, and thoroughly believing that this wild stranger from the sea might well turn cannibal, the dog-drivers headed north-west toward Tomat, the solitary village on that northern Arctic coast. For three days our labouring dogs dragged us through the drifts along the road to Tomat, fortunately for us following a chain of deserted huts in each of which we found refuse scraps of fish-heads, entrails of reindeer, and such similar offal, the which we (both men and dogs) ate greedily to save us from starvation, and on the fourth day, so frozen that I had to be carried from my sledge into a hut, we arrived at Tomat.

Staying there only a day to thaw out, to change my dogs for fresh ones, and to replenish my food supply (in that poor village, itself facing the winter with scanty food, getting each solitary fish was harder indeed than extracting from the villagers their teeth), I started east along the Arctic coast, with my feet so badly frozen I could not walk.

By evening, marked by a pole, I found the cache De Long had

left on the beach, but so thick was the falling snow I could not see the first cutter offshore. Salvaging the log-books and the *Jeannette's* navigating outfit, I loaded them on my sledges and turned south till I came on the first hut where De Long had stopped. For a week after, amid frigid Arctic gales with the temperature far below zero, I searched along the solidly frozen Lena, visiting every hut, finishing finally in that hut on the promontory where for three days De Long had waited for the rivers to freeze so he might cross, and where Dr. Ambler had sliced off Erichsen's toes. There beyond the frozen river, on the wind-swept farther shore, for a short distance I could follow where his toiling shipmates had dragged Erichsen along on his sledge, for the deep grooves left in the soft slush a month and a half before now stood clearly out in solid ice.

But there finally I lost the trail. The deep drifts of many snows buried all tracks. Facing a myriad of wandering streams, any one of which De Long might have followed south, I searched in vain for further tracks, for the hut in which Erichsen had finally died, for the epitaph-board which Nindemann told me he had left there to mark it, but not another trace of De Long or of his party could I find in the ever-thickening snow as storm succeeded storm and buried the Lena Delta in drifts so deep that my floundering dogs could scarcely drag me through them.

It was now late November, six wintry weeks since without food and without shelter, De Long had parted somewhere thereabouts in that ghastly wilderness from his two messengers. Only one of two things now was possible—either De Long and his party had somehow been found by natives who were sheltering him, quite as safe as I myself; or he had long since perished and was somewhere buried beneath the snowdrifts on the open tundra, where in the dead of winter it was hopeless to search for him. Weak and frozen myself from my desperate search, coming on top of my long exposure in the open whale-boat, it was now imperative that I get out of the delta before my frozen corpse found an unmarked grave beneath the snows alongside my missing shipmates. So sadly I ordered my worn dogs south. It took us a week to fight our way back to Ku Mark Surk at the delta head, and two days more to cover the final fearful miles along the Lena through the mountain gorges up to Bulun, where at last at the end of November I arrived, sick at heart at my failure to find my comrades, terribly sick physically from rotten food, from hunger, and with numbed limbs from which the Arctic cold had drained away every vestige of life.

CHAPTER XXXVIII

ALL winter long, while endeavouring to recuperate my frozen arms and legs, I gathered supplies and sledges from Bulun, from far-off Yakutsk, from all the villages between, for an intensive search of the delta in early March before the annual spring-time freshets, feeding the Lena with the melting snows of southern Siberia, should come pouring out on the flat delta, burying it in a flood of raging waters and sweeping my shipmates out into the Arctic Sea.

I kept only Nindemann, now recovered, and Bartlett with me to help me in my search. All the remaining survivors, a pitiful party, under Lieutenant Danenhower's charge, went south over the fifteen-hundred-mile track to Irkutsk. Poor Aneguin, weakened by exposure, died before he got out of Siberia; Jack Cole, violently insane, reached America only to die soon after in a government asylum; and Danenhower himself, broken in health, after a few brief years spent undergoing a long series of operations, soon followed him to the grave. The rest, except for Leach, whose toes had to be amputated, reached America safe and sound. Meanwhile by courier from Bulun to Irkutsk, the head of the telegraph lines in Siberia, the news of the disaster to the *Jeannette* finally went out on December 21.

For two and a half years not a word of us had ever reached civilisation. As the months since our departure lengthened into years and no news came, anxiety in America and in Europe over our fate deepened into keen alarm. Swallowed by the trackless Arctic, fear for us grew, and in the summer of 1881, two relief expeditions fitted out by the American Government went north to search for us. But where should they look? Which way did the polar currents go from Behring Strait where we had entered? No one knew save we on the *Jeannette,* and our knowledge was useless to a world facing a search of the unknown north.

One expedition in the Revenue Cutter *Corwin* searched for us fruitlessly off Wrangel Land, but not daring to enter the ice, found no trace. A second expedition, in the U.S.S. *Alliance,* thinking perhaps we might have drifted east over North America and come out beyond Greenland into the Atlantic, searched during the whole summer the fringe of the polar pack around Spitzbergen, getting in open water as far as 82° North, five degrees higher than we in the *Jeannette* were ever carried by the pack before it crushed us.

But neither expedition found the slightest sign of us, and more alarmed than ever, an international search was being organised by our Navy, with the help of England, Russia, and Sweden for the summer of 1882. In the midst of these preparations in late December, 1881, from far up in the Arctic Circle, my first brief telegram from Bulun at last reached Irkutsk and flashed out over the wires to an astonished world, ending the mystery of the *Jeannette's* disappearance, bringing joy to some whose friends had definitely escaped; blank despair to others whose lives were bound up with poor Chipp and his lost boat's crew; and a terrible state of mingled fear and hope, not to be resolved for unknown months yet, to Emma De Long and the families of those men still with her husband. I felt that they were dead, but I did not know it, and dared not say so. I could only announce them as having landed safely, but yet unfound. My heart ached for Emma De Long, half the globe away from me, clinging to her daughter, praying that her husband might yet be alive, tortured by the long-drawn-out fear of waiting for word from Siberia, dreading each knock at the door as announcing the messenger bringing definitely the black news of his death, and all the while with her imagination able to dwell only on the agonies which her husband had undergone, and if by some miracle (for which she prayed) he still were living among those Arctic wastes, he must yet be suffering.

I received *carte blanche* from Washington for funds to pursue the search; from St. Petersburg, I was assured all the resources of Russia were at my command. But Washington and St. Petersburg were far away from the trackless delta where I must pursue my search, and *carte blanche* telegrams helped me little. A few dogs, a few interpreters, a supply of dried fish sold under compulsion by natives who could ill afford to spare them, was the total extent of the assistance I could use and get delivered to distant Bulun up in the Arctic Circle, fifteen hundred long miles away from civilisation and the telegraph-wires at Irkutsk, when in late February with practically all the fish in the Lena Delta in my possession and the poor Yakuts face to face with famine, I resumed my search.

Dividing my forces, I sent Bartlett and an interpreter to cover the eastern branches of the Lena, while with Nindemann to guide me, I started again to search the western branches myself.

I had seven dog teams hauling fish, having practically stripped Jamaveloch and every Lena village of its entire supply. Delayed considerably still by fierce snow-storms, we went north from Ku Mark Surk into the delta, but it took two weeks for the straining dogs to drag our stores along to where the Lena started branching widely at Cass Carta, and many a burdened dog froze to death in the drifts before he got there. At Cass Carta at last, I reorganised my remaining teams, and on March 12, still in the midst of winter weather, sent Bartlett east, and with Nindemann, began myself the search of western rivers.

For a week, systematically Nindemann scouted along each river, trying to pick out the one that he and Noros had followed south. But the innumerable storms since had changed the whole face of that frozen country. How many streams we examined, I cannot even guess. Nindemann, his broad brows knit with puzzled furrows, could find nothing familiar in any of them. Baffled, we gave up searching there and went far to the north, to follow down De Long's trail from the coast, but at the same point where in November I had lost the track, Nindemann himself was able to do no better in pointing out the path. And then came a raging storm which held us snow-bound for three days.

Despairingly, I considered the situation. Would we ever pick up De Long's track? It must be soon or never! Before long the river ice would break up, we could no longer travel, and swollen with melting snow from the whole interior of Siberia, the Lena would come flooding down in torrents to drown out the low delta lands, washing away for ever every trace of my comrades! De Long must be somewhere to the south. In desperation, I gave up searching the central delta for his track, and decided to go back again to the delta head, to sweep the spreading rivers there as I came north.

Soon after, starting from the southward again, since Nindemann also felt that there he could do best, we began at a wide bay, from which one tremendous river flowed eastward toward Jamaveloch, another flowed westward and northward toward Tomat, and in between the Lena, in many smaller branches, flowed due north, spreading widely out and meandering over the delta, though now, of course, it flowed beneath the ice as every stream was still solidly frozen over.

Following the edge of this tremendous bay, I examined every headland on it. Broken slabs of ice were piled up in tangled masses on the banks; the snow, drifted by the winds, ran in smooth slopes from the river ice to the tops of the promontories, filling in the banks; dozens of frozen streams, like twigs spreading from a limb, branched out from the bay, complicating the search.

Coming in the late afternoon to a high headland on the western side of the bay, I left my sledge as usual on the river ice, and clambered up the crust of snow to its top. The crest was strongly wind-swept and fairly bare of snow; as I stooped to brace myself against the wind, I saw right on the point of the promontory signs of a long-dead fire, with half-burned driftwood logs hove into the wide bed of ashes and apparently many footprints in the ice about.

Beckoning to Nindemann to come up, I asked him:

"Did you or Noros build that fire here?"

"No," said Nindemann, "it looks to me we came this way, but we never had a fire like that."

I motioned up my dog-driver, questioned him in native dialect:

"Do Yakuts build fires this way?"

"No, no, master," he protested volubly, "Yakuts build only small fires, never big fires like this."

"Well, Nindemann," I said, "I think we're on the trail at last. This looks to me like a signal fire, especially since it's built on this promontory to shine out over the bay. De Long must have passed here."

"Yah," agreed the quartermaster, "that is right. There! See? There is the old wreck of a flat-boat on the bank and I remember Louis and me passed by that wreck the same day we left the captain! This is the way we came, and the captain said he'd follow our trail!"

Going down to the river again, we climbed aboard our dog sledges. Nindemann on his sledge led along the ice, and with me following on mine, we set off on a short journey up the stream to examine the bare skeleton of that flat-boat, stranded on the bank a quarter of a mile down-stream.

I rode, sitting sideways on my sledge, facing the high bank which rose some thirty feet above the river, and which, as usual, had hard-driven snow packed in a glistening slope from its crest down over the frozen river. Going swiftly along over the ice this way while eagerly scanning the river-bank, I noted standing up through the sloping snow what seemed to be the points of four sticks lashed together with a rope.

Immediately I rolled off the speeding sledge, and swiftly going to the spot, found a Remington rifle slung from the sticks, its muzzle some eight inches out of the snow. A real sign of De Long at last!

Instantly I sent my driver to bring Nindemann back, feeling that here the weakening wanderers might have made a cache of such belongings as they could no longer carry, and perhaps even have left a record of their progress. We were certainly on the trail now!

While the Yakuts at my orders began digging in the snow around those sticks, Nindemann returned to the flat-boat, and I with a compass again climbed up the steep river-bank, intending to get some bearings from which later I might find that spot in case a sudden snow-storm should blot out the way to it.

Panting from my exertions, I looked about for a good place on that high ground from which to take the compass bearings when a few steps off, partially buried in the snow still left on that forlorn and gale-swept height, I saw a copper tea-kettle. With a beating heart, I started for it, then stopped short. There before me on that

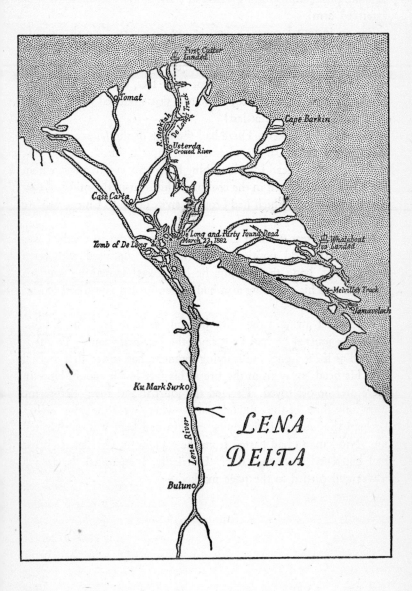

desolate plateau, protruding stiff and stark above the snow—was an extended arm!

For an instant I gazed, aghast at my discovery, then dropped to my knees to find that that arm belonged to Captain De Long! There he lay, cold and silent in death, half-buried in the snow. A yard or two off lay Dr. Ambler, while near their feet, closest to where the fire had been beneath the copper kettle, was stretched Ah Sam. My long search was ended at last!

Mournfully I looked. There had the saga of the *Jeannette* ended, there in the Arctic snows was my lost captain—dead. For a long time with bowed head I knelt sobbing before my commander, whom last I had seen, erected in the cockpit of his boat in the midst of that roaring polar gale which had brought swift death to Chipp, waving me on to safety.

As I gazed tear-stricken into his face, calm even in death, I was struck by the odd position of his left arm, upraised with open fingers as if, lying there dying, he had tossed something over his shoulder and his stiffening arm had frozen in that gesture. I looked behind him.

A few feet away in the snow beyond his head lay a small note-book, the journal he had kept since the *Jeannette* sank. To me it seemed as if De Long, in his dying moment, had tossed that journal over his head, away from the fire at his feet lest it should blow in there and be destroyed. I seized the journal and rose. Before me were only three of the captain's party—where were the other eight? Perhaps the journal, if the dying captain had kept it up, might tell me. Nindemann had parted from the captain on October 9. What had happened since that day? Hurriedly, I separated the frozen leaves and turned to the page marked——

CHAPTER XXXIX

OCTOBER 10, Monday—120th day.
Last half-ounce alcohol at 5.30; at 6.30 send Alexey off to look for ptarmigan. Eat deerskin scraps. Yesterday morning ate my deerskin foot-nips. Light S.S.E. airs. Not very cold. Under way at eight. In crossing creek three of us got wet. Built fire and dried out. Ahead again until eleven. Used up. Built fire. Made a drink out of the tea-leaves from alcohol bottle. On again at noon. Fresh S.S.W. wind, drifting snow, very hard going. Lee begging to be left. Some little beach, and then stretches of high bank. Ptarmigan tracks plentiful. Following Nindemann's tracks. At three halted, used up; crawled into a hole in the bank, collected wood and built fire. Alexey away in quest of game. Nothing for supper except a spoonful of glycerine. All hands weak and feeble, but cheerful. God help us.

October 11, Tuesday—121st day.
S.W. gale with snow. Unable to move. No game. One spoonful of glycerine and hot water for food. No wood in our vicinity.

October 12, Wednesday—122nd day.
Breakfast; last spoonful of glycerine and hot water. For dinner, we tried a couple of handfuls of Arctic willow in a pot of water and drank the infusion. Everybody getting weaker and weaker. Hardly strength to get firewood. S.W. gale with snow.

October 13, Thursday—123rd day.
Willow tea. Strong S.W. wind. No news from Nindemann. We are in the hands of God, and unless He intervenes, we are lost. We cannot move against the wind, and staying here means starvation. Afternoon went ahead for a mile, crossing either another river or a bend in the big one. After crossing missed Lee. Went down in a

hole in the bank and camped. Sent back for Lee. He had turned back, lain down, and was waiting to die. All united in saying Lord's Prayer and Creed after supper. Living gale of wind. Horrible night.

October 14, Friday—124th day.
Breakfast, willow tea. Dinner, one half-teaspoonful sweet oil and willow tea. Alexey shot one ptarmigan; had soup. S.W. wind moderating.

October 15, Saturday—125th day.
Breakfast, willow tea and two old boots. Conclude to move on at sunrise. Alexey breaks down, also Lee. Come to empty grain raft. Halt and camp. Signs of smoke at twilight to southward.

October 16, Sunday—126th day.
Alexey broken down. Divine Service.

October 17, Monday—127th day.
Alexey dying. Doctor baptized him. Read prayers for sick. Mr. Collins's birthday; forty years old. About sunset Alexey died; exhaustion from starvation. Covered him with ensign and laid him in the crib.

October 18, Tuesday—128th day.
Calm and mild, snow falling. Buried Alexey in the afternoon. Laid him on the ice of the river and covered him with slabs of ice.

October 19, Wednesday—129th day.
Cutting up tent to make footgear. Doctor went ahead to find new camp. Shifted by dark.

October 20, Thursday—130th day.
Bright and sunny, but very cold. Lee and Kaack done up.

October 21, Friday—131st day.
Kaack was found dead about midnight between the Doctor and

myself. Lee died about noon. Read prayers for sick when he found he was going.

October 22, Saturday—132nd day.
Too weak to carry the bodies of Lee and Kaack out on the ice. The Doctor, Collins and I carried them around the corner out of sight. Then my eye closed up.

October 23, Sunday—133rd day.
Everybody pretty weak. Slept or rested all day and then managed to get enough wood in before dark. Read part of Divine Service. Suffering in our feet. No footgear.

October 24, Monday—134th day.
A hard night.

October 25, Tuesday—135th day.

October 26, Wednesday—136th day.

October 27, Thursday—137th day.
Iversen broken down.

October 28, Friday—138th day.
Iversen died during early morning.

Oct. 29, Saturday—139th day.
Dressler died during night.

Oct. 30, Sunday, 140th day.
Boyd and Görtz died during night. Mr. Collins dying.

CHAPTER XL

MR. COLLINS dying.

And there on October 30, the pitiful record ended. Before he could put a period to that final tragic sentence, the pencil dropped from De Long's nerveless fingers, with his last conscious effort he tossed his journal over his shoulder to save that record of what had happened to his shipmates from the fire near-by. My blurred eyes stared at the pages before me; my captain had died as he had lived—with his thoughts only on his men. Not a word on that last tragic page about himself, his sufferings, or his own approaching death. And yet for the solitary malcontent who on the *Jeannette* had tried the captain's very soul, who had fought savagely to destroy the discipline on which De Long relied to save our health and our lives amid the perils of the Arctic pack, he could still tax the little strength left in his starved body to note down:

"Mr. Collins's birthday; forty years old."

And so to the final entry, by the irony of Fate recording that man's death, the dying captain's stiffening fingers scrawled out faithfully the record of his shipmates, but not one word regarding George Washington De Long!

With wet cheeks, I stood humbly before the frozen body from which the great soul of my Captain had passed, till finally Nindemann approached, and with his aid I loosened the three terribly emaciated forms from the snow and bore them gently to our sledges. Evening fell and we returned to a hut at Mat Vay, ten miles south across the great frozen bay, where I established headquarters. The next few days, digging in the snow near the tripod where I had found the rifle, we uncovered the bodies of the rest of the party, all fearfully gaunt. There was not a whole garment on any man; and not one pair of boots or of fur clothing could we find. Everything made of skin or of leather had been eaten; most of the men lay in ragged underwear with their feet bound in canvas, and the first two who died had been stripped naked and so lay in the snow, their poor rags wrapped round their then dying comrades.

As we dug away the snow in the lee of that river-bank where the last ten survivors (Alexey had died a short distance away near that grounded raft) had huddled, trying to shelter themselves from the fierce gales, we found first the ashes of their fire, then the sticks with which they had sought to rig their sole remaining piece of canvas as a wind-break, and then so close to the ashes that their underwear was badly scorched, the bodies of all hands except Lee and Kaack. And there also we found De Long's main ship journals. But those two men and the expedition's silken ensign we could not find.

I puzzled over that, and then reading back again De Long's journals, I noted that Lee and Kaack had been carried "around the corner out of sight." But where was there a corner in that bank running straight north and south? And then it came to me that as all the gales De Long had logged blew from the southward, they must have set their bit of canvas up athwart the wind and camped on its north side, so that he meant around the corner *of the tent*. Directing the natives to dig to the southward of the sticks, they soon found Lee and Kaack, naked both, and now there was nothing missing but the ensign. Knowing well that De Long, however weak he might be, would never have abandoned that, I ordered the edge of the tent line excavated, and there at last we found the silken banner, deep in the snow, safely rolled in its oilskin case.

But one thing still puzzled me. Why were the men whose deaths De Long had recorded all there in the lee of that high bank, while he himself, with Ah Sam and Dr. Ambler, the last survivors, lay on top of that promontory where there was not the slightest shelter from the biting wind? After another survey, I could only conclude that De Long, wholly despairing of rescue and feeling death swiftly approaching, had with his two dying companions started to move the records of the expedition up from the river-bank on to the higher ground where they would longer escape the spring floods, but the three of them having made one trip up the slope in which they dragged with them the copper kettle and a tin chart-case (which I found there near the captain) had none of them the strength to crawl back for another load, and there they all soon perished. Evidently of those three Ah Sam died first; his arms were crossed above his breast as if laid out by the others close to the little fire they had built beneath the kettle in which they were trying to boil a few twigs of Arctic willow. Whether Dr. Ambler or Captain De Long was the last survivor, no one will ever know—Ambler lay face down near the fire, De Long a little farther off.

At my direction, Nindemann and Bartlett carefully searched the camp and all the bodies for any final messages left, but only on Surgeon Ambler did they come across anything like that. The last page of his journal was in the form of a letter. I sobbed as I read it.

On The Lena,
Thursday, Oct. 20, 1881.

To Edward Ambler, Esq.,
    Markham P. O., Fauquier Co., Va.
My dear Brother:

I write these lines in the faint hope that by God's merciful providence they may reach you at home. I have myself very little hope of surviving. We have been without food for nearly two weeks, with the exception of four ptarmigans amongst eleven of us. We are growing weaker, and for more than a week have had no food. We can barely manage to get wood enough now to keep warm, and in a day or two that will be passed. I write to you all, my mother, sister, brother Cary and his wife and family, to assure you of the deep love I now and have always borne you. If it had been God's will for me to have seen you all again I had hoped to have enjoyed the peace of home-living once more. My mother knows how my heart has been bound to hers since my earliest years. God bless her on earth and prolong her life in peace and comfort. May His blessing rest upon you all. As for myself, I am resigned, and bow my head in submission to the Divine will. My love to my sister and brother Cary; God's blessing on them and you. To all my friends and relatives a long farewell. Let the Howards know I thought of them to the last, and let Mrs. Pegram also know that she and her nieces were continually in my thought.

God in His infinite mercy grant that these lines may reach you. I write them in full faith and confidence in help of our Lord Jesus Christ.

Your loving brother,

J. M. AMBLER.

CHAPTER XLI

ATOP a rocky promontory looking to the north, towering four hundred feet above the great bay of the Lena Delta and far beyond the reach of any possible flood, I prepared for my captain and his crew their final resting-place. Excavating from the solid rock a foundation, I built in the form of a huge cairn, a monumental rectangular stone structure visible easily twenty miles in all directions, making its sides of the thick planking torn from that wrecked flat-boat near the last fatal camp, and covering the stout planking with rough stone quarried on the mountain-top. Above that rocky cairn, I raised a massive cross twenty-five feet high, hewn from a driftwood spar salvaged from the bay below, and upon the spreading arms of that cross, I cut the names of those who were to rest beneath it.

When all was ready, on April 6, 1882, on that gale-swept mountain-top overlooking the Lena, we buried them. Composed wholly of sledges, the long funeral procession of straining dog-teams wound across the snow-covered tundra and up the ice-coated slopes of that mountain, the dark sledges bearing the silent seamen standing starkly out against the whiteness of the driven snow, with the one bit of colour there, the *Jeannette's* silken ensign draping the cold figure of her captain. On foot the three survivors present, Bartlett, Nindemann, and I, trudged sadly along. Arrived at the cairn, we three lifted the thin bodies from the sledges, tenderly laid them out on a bed of snow inside the tomb, Captain De Long at one end, then the others in order of rank: Surgeon Ambler, Mr. Collins, Lee, Kaack, Görtz, Boyd, Iversen, Dressler, and last at the other end, Ah Sam. Then reverently removing the ensign from the captain's body that I might return it to her hands who fashioned it, we took our long last look at our dead comrades.

In that deep Arctic solitude with no unhallowed lips droning out unfelt phrases, we who had lived with them in toil and peril and nearly died with them in anguish, stood with bowed shoulders and bared heads in the freezing wind before our dead, and with choking voices murmured our heart-felt farewell:

"Good-bye! Sleep well, shipmates!"

And then sorrowfully sealing up the cairn, we left them to their rest. Never had heroic explorers a more fitting tomb. Amidst the Siberian snows, looking out over the Lena's great bay at the desolate cape below which had witnessed their last agony, and northward across that Polar Sea which he had valiantly given his life to conquer, De Long and his men of the *Jeannette* lay at last beneath the huge cross on that rocky cairn, with the fierce Arctic gales they had so often bravely faced mournfully wailing their eternal dirge.

EPILOGUE

I AM an old man now, looking forward soon to joining those shipmates of long ago. Many honours have come my way, and for sixteen years as an admiral in the Navy, I have had charge of designing and building the throbbing engines which drive our every warship. That mighty fleet which so proudly showed the American flag in every ocean on the globe and has just returned to confound the doubters at home and abroad who foresaw those ships with broken-down machinery cluttering every port from Rio to Yokohama, was my design. The engines and the boilers which so sturdily drove the *Oregon* twelve thousand miles around South America under forced draught in time to take her place in the forefront of the battle-line at Santiago and deal the death blows to Spanish sea power, were my creation.

From that day in 1861, when as a young engineer officer I joined the Navy, until the day forty-two years later when I retired full of honours as its Engineer-in-Chief, machinery had been my life, and I had hoped that my name might as a result find its place with that of Ericsson as one who had done much to advance the application of power to our warships. But as the years since my retirement weigh me down, and I see my proud *Oregon* already vanished from the fleet, and that fleet itself ere long destined to disappear before the creations of newer and better engineers than I, more and more do I realise that it is the men themselves and how they lived and died, rather than their puny handiwork, which those who come after us will ever have reason to cherish as the true measure of any man.

And so that huge cross I reared in the Lena Delta amidst the polar snows looms larger and larger in my mind, and now I only humbly hope, as approaching the end of my long days I look back over my life, that the name of George Wallace Melville may be a little remembered as one of those who served on that far-off cruise in the *Jeannette,* when my science and machinery faded from all importance, engineering went wholly by the board, when first with only our stout ship to shield us and then without her, face to face with Nature in her fiercest mood for endless months we battled the Arctic ice beneath the banner of George Washington De Long, and in his life and death I learned what truly makes the man.